Schriftenreihe des Energie-Forschungszentrums Niedersachsen (EFZN)

Band 12

Das EFZN ist eine wissenschaftliche Einrichtung
der Technischen Universität Clausthal in
Kooperation mit den Universitäten Braunschweig,
Göttingen, Hannover und Oldenburg.

efzn
Energie-Forschungszentrum Niedersachsen

TU Clausthal

Prediction of Pressure and Temperature in CO_2 Injection Wells Based on Analytical Modeling

Dissertation

Submitted in partial fulfillment of the requirements for the award of the degree of

Doctor of Engineering

Of the Faculty of Energy and Management Sciences

Of the Clausthal University of Technology (TU-Clausthal)

Germany

Dy

Arron A. Tchouka Singhe

Clausthal-Zellerfeld, 2012

Bibliografische Information der Deutschen Nationalbibliothek

Die Deutsche Nationalbibliothek verzeichnet diese Publikation in der Deutschen Nationalbibliografie; detaillierte bibliografische Daten sind im Internet über http://dnb.d-nb.de abrufbar.

1. Aufl. - Göttingen : Cuvillier, 2013

Zugl.: (TU) Clausthal, Univ., Diss., 2012

978-3-95404-405-4

Chairperson of the Examination Board
Prof. Dr. Norbert Meyer
TU-Clausthal

Chief Reviewer
Prof. Dr.-mont. Leonhard Ganzer
TU-Clausthal

Reviewer
Prof. Dr. -mont.I.R. Günter Pusch
TU-Clausthal

Reviewer
Prof. Jan Rune Ursin
University of Stavanger, Norway

Nonnenstieg 8, 37075 Göttingen

Telefon: 0551-54724-0

Telefax: 0551-54724-21

www.cuvillier.de

1. Auflage, 2013

Gedruckt auf säurefreiem Papier

978-3-95404-405-4

SUMMARY

Injection into geological formations is seen by many as a short to medium term measure to reduce emissions of CO_2 to the environment and as such to slow down the pace of global warming. The injection process requires that the fluid flows effectively into the host formation. To this end it is very important to be able to predict the pressure and temperature at the bottom of the well. The bottomhole pressure and temperature constitute with the reservoir pressure and reservoir properties, the main parameters that govern the rate at which CO_2 flows into the storage reservoir. As a consequence, these parameters would also influence the major physical and chemical processes occurring in the near wellbore region. In this light the determination of the injection rate and the prediction of the near wellbore physicochemical phenomena, require accurate prediction of the bottomhole pressure and temperature as a function of wellhead pressure and temperature. This would require prediction of pressure and temperature profiles along the well to identify possibility of phase transition and abrupt densities changes which often make the bottomhole pressure and temperature to become difficult to estimate.

In the petroleum industry wellbore flow modeling and simulation is often used to predict fluid temperature and pressure profile along a well. The existing codes are developed for generic petroleum reservoir fluids. For these fluids high accuracy in the estimation of temperature along the well is not always very important. This is justified by the small dependency of the pressure of petroleum fluids on temperature. Moreover in the well these fluids generally remain either in the liquid phase (water, oil) or in the supercritical phase (gas). This is not the case for CO_2 that has critical temperature around 31.04 °C and therefore is likely to experience phase transition along the well. Phase transition renders temperature and pressure estimation difficult and may create problems of flow stability and assurance. To ensure a stable and easily predictable flow, it has been custom to inject CO_2 in its supercritical form, where it is known to flow more effectively. Some costly measures would be also adopted to insulate the upper parts of the injection wells in order to prevent cooling which would induce phase transition.

A consequence of not rigorously estimating the fluid temperature along the well is the negligence of the dependency of the temperature of the fluid on injection time. In fact as injection goes on with time, heat is exchanged between the fluid in the well and the rock surrounding the wellbore. This exchange of heat alters the thermal equilibrium that initially existed between the well and the rock. As such it significantly changes the rate at which and eventually also the direction in which heat is transferred between the surrounding rock and the fluid. This will therefore influence the temperature profile of the fluid in the well and eventually result in a dramatic change in the density profile and hence in the pressure profile.

Another issue that could deter the accuracy with which existing codes predict pressure and temperature along the well is that these codes generally rely on conventional 3-parametric cubic equations of state and correlations to estimate the fluids' thermodynamic and transport properties respectively. These equations and correlations are known to be of unsatisfactory accuracy, especially in the liquid and near critical fluid regions. As a consequence the estimation of thermodynamic and transport properties would be with large errors that will be inherited by pressure and temperature. Furthermore, the solutions to the pressure and temperature equations in these codes are obtained with numerical approximations. These approximations would add errors to the predicted values of temperature and pressure.

The forgone points justify the need for innovative and robust models and solutions that will not only accurately estimate temperature and pressure along the well as a function of depth and time, but will also predict phase changes and their effects on temperature and pressure. This work addresses this need through appropriate models and analytical solutions for predicting pressure and temperature along the well. One pressure model and four temperature models have been developed.

The main temperature model is the most rigorous in estimating the temperature along the well. It is non-adiabatic and as such it takes into account the temperature effects caused by compression/expansion of the fluid in the well and those related to the heat exchanged between the fluid and the surrounding rock. In this way the wellbore and the rock are assumed to form a heat exchanger with composite wall between the heat source and the heat sink. The composite wall is made up of the wellbore components (cement, casing, annulus and tubing).The three other temperature models are the adiabatic model, the geothermal model and the isothermal model.

The adiabatic model assumes that the well is perfectly insulated from the surrounding rock. Consequently only the effect of compression and/or expansion influences the temperature of the fluid in well. The geothermal model assumes that the temperature of the fluid is equal to that of the surrounding rock at each depth. This situation would be approached whenever the flow rate is very low to allow longer contact time between the fluid and tubing surface. The isothermal model assumes that the temperature is constant along the well. It would be approached in a horizontal well when inlet temperature is equal to geothermal temperature and frictional pressures losses are sufficiently small not to cause significant temperature drop.

The non-adiabatic temperature model associated with the pressure model constitute a building block for rigorous prediction of pressure and temperature in a gas injection well of any configuration and deviation angle, and in layered formations with differing thermal conductivities.

Beside the use of analytical solutions to pressure and temperature, the accuracy of the predictions is further guaranteed by the implementation of highly accurate fundamental equations of state to estimate thermodynamic and transport properties of CO_2. The equation of Span and Wagner, the equation of Fenghour et al. and the equation of Scalabrin et al. are used for the estimation of thermodynamic properties, viscosity and thermal conductivity respectively.

The limitations of the equations of state to CO_2 makes it that this work is also restricted to CO_2 only. However it can be extended to any other gas and mixture of gases by using appropriate equations of state.

The transparency and control of the models' performance is ensured by the use of dimensionless numbers. These numbers also ease analysis of the results and detection of user-related errors.

The major innovations can be summarized in the following points:

1. Development of a workflow for using high accuracy multiparameters equations of state to predict CO_2 thermodynamic and transport properties along the well.
2. Revisiting of existing procedure to calculate the overall heat transfer coefficient between the rock and the fluid
3. Adapting an innovative analytical model developed for temperature prediction during well testing to predict temperature propagation in the rock at any injection time, being it early or late.
4. Expressing pressure and temperature equations in terms of dimensionless numbers in order to enable easiness and transparency of the solutions and ease control of the codes' performance, results analysis and detection of user-related errors
5. Revisiting the marching algorithm and implementing it with analytical solutions to arrive at accurate estimation of pressure and temperature along the well.
6. Development of a computer code that accommodates wellbores in layered formations with different thermal properties and varying geothermal gradients based on analytical models.

The solution to the models has been implemented through simulation codes developed with the Visual Basic.Net programming language on the Microsoft Visual Studio platform. The codes are encapsulated in a user-friendly graphical user interface. This interface ensures easy input and output of data.

To ensure fast computation and preserve good accuracy at the same time, tables of thermodynamic and transport properties have been created in the temperature and pressure ranges relevant for CO_2 injection into geological formations (temperature between 250 K and 450 K and pressure between atmospheric to 400 bar). The codes read these properties at prevailing temperature and pressure in the wellbore and where applicable, they interpolate for values falling between those tabulated.

The marching algorithm is implemented to minimize the errors related to the use of average values of thermodynamic and transport properties for analytical solutions of pressure and temperature in the well. This involves the division of the wellbore into segments of which lengths are determined by some validation criterion that ensures high accuracy of the solution. This criterion requires that the relative difference between the z-factor at the inlet of a segment and that at its outlet should not exceed 0.1%. A consequence of the segmentation of the well is the establishment of a computational loop which is repeated from the inlet to the outlet of the pipe.

The predictions of the codes have been compared with measured data from two CO_2 injection projects in saline aquifers under different injection conditions. The first project is the Ketzin project. It is an onshore project in the north of Germany. Eight straight injection days with almost similar wellhead conditions (temperature, pressure, and flow rate) have been selected from the dataset. CO_2 is injected at low (~30000 m^3/day) rate and flows through a shallow vertical well to the aquifer reservoir at 630 meters depth. The flow takes place mostly in the gas and gas-like supercritical phases. The second project is Snøhvit project in Norway. In this project CO_2 has been injected through a deviated offshore well of around 2600 meters true vertical depth. Eleven continuous injection days under similar wellhead conditions have been selected from the dataset. CO_2 flows at a relatively high rate (~500000 m^3/day) in the liquid and liquid-like supercritical phases. For these two cases the relative deviation of simulated values from measured values are between 0.01% and 1.93 % for pressures while for temperatures they are between 0.33% and 10.83%.

Thermal conductivities of the surrounding rock have been the major matching parameter for temperature in the well. The best match for the Ketzin well has been obtained by considering that the surrounding rock has three layers with different thermal conductivities, 3.0 W/mK, 2.5 W/mK and 2.0 W/mK respectively. These thermal conductivities values are within the range reported for a reservoir in the Altensalzwedel area which is in the North German Basin like Ketzin. The best match was obtained for Snøhvit by assuming one rock layer with same thermal conductivity equal to 1.25 W/mK. The lower thermal conductivity in one rock layer is in line with the fact that Snøhvit well is offshore, below water and may be surrounded by rock of fairly consolidated sand with high porosity and high water content.

The matching process indicated that the temperatures of the fluid and surrounding rock are very sensitive to the rock thermal conductivities. This point has also been confirmed in this work through the analysis of the sensitivity of fluid and rock temperatures to rock thermal conductivity. It further corroborates the fact this work can be used to estimate thermal conductivity of the rock whenever measured values of temperature of the fluid in the well are available.

Further examination of the accuracy of this work has been conducted by comparing its predictions with measured data from a CO_2 production well. The well has a long section deviated at 26.5°. The results are also very accurate. The maximum deviations are 3% and 4% for pressure and temperature respectively. It is however noteworthy that a general implementation of the code on production wells would require accounting for slip between phases in case of high liquid content. Phase slip has been neglected in this work.

More insight in the agreement of the models with theory has been obtained by comparing the predictions of the non-adiabatic temperature model with those of the other earlier mentioned temperature models. The conclusions are that this work agrees with theoretical principles of fluid flow into cylindrical pipes and those of heat transfer in a heat exchanger with a composite wall and in solids. Further confirmation of the agreement with theory has been also obtained through analyzing the sensitivity of the predictions to injection parameters (wellhead temperature, wellhead pressure, flow rate and tubing inner diameter and inner surface roughness) and injection time.

While acknowledging the high accuracy of the procedure developed in this work in predicting temperature and pressure of CO_2 in an injection well, it is also important to recognize the room that exist for improvements in many aspects. These improvements are closely related to the assumptions underpinning the temperature and pressure models.

One of these assumptions is very conservative and considers that the vertical heat propagation in the surrounding rock and in the fluid is to be neglected. Another one that could have some sensitive implications on the predictions is that the heat of vaporization and the heat liberated upon condensation in case of phase transition are assumed to propagate all into the formation through the tubing over the entire length of the well. This is not totally true because a part of this heat will be carried in the fluid by convection. Furthermore the mass exchange between phases when the gas phase encounters the liquid phase in the well is not taken into consideration. Consequently the models will not be able to simulate the temperature and pressure profile in a shut-in well experiencing boiling and condensation with acceptable accuracy. This is of importance because shut-in situations will always occur during real field injection.

In the light of the foregoing it is recommended that, further work would consist of modeling the mass transfer and heat transfer phenomena with analogy to the phenomena at play in a distillation column. It would also be important to take into consideration the vertical propagation of the temperature in the rock. Another further work, which could also significantly improve the accuracy of the predictions with this work would consist of directly implementing the multi-parameter equations of state in the codes by taking advantage of the high speed of today's and future's processors to develop time-efficient codes with these equations. The codes would also be generalized for other gases or mixtures of gases by including tables of properties generated with dedicated multiparameters equations of state or by directly implementing these equations in the codes. Including phase slip through drift-flux models for instance would enable to also use the codes for prediction of temperature and pressure in production wells producing gases with high liquid content.

As final conclusion it can be said that although this work has room for improvements, it has shown to be very reliable in predicting temperature and pressure of CO_2 in an injection well as function of depth and time. It has also shown to accurately predict temperature and pressure in a CO_2 production well with low liquid content where phase slip is negligible.

Kurzfassung

Die Speicherung von Kohlendioxid in geologischen Formationen wird von vielen als eine (Brückentechnologie) im kurz- bis mittelfristigen Bereich gesehen, um eine Verlangsamung der globalen Erwärmung zu erreichen.

Der Prozess der Injektion erfordert eine effiziente Strömung(hohe Permeabilität) in der Zielformation. Um dies zu erreichen, ist die Vorhersage der Bohrlochtemperaturen sehr wichtig. Die Temperaturen und Drücke am Bohrlochboden korrelieren mit dem Lagerstättendruck und mit weiteren Lagerstättenparametern, die wiederum die Fließgeschwindigkeit des Kohlendioxids beeinflussen.

In Folge dessen lassen diese Parameter Rückschlüsse auf grundlegende physikalische und chemische Prozesse des näheren Bohrlochumfeldes zu.

In diesem Licht sind die Bestimmung der Injektionsrate und die Vorhersage des näheren Bohrlochbereiches der physikalisch-chemischen Phänomene erfordern eine präzise Vorhersage von Bohrlochbodentemperatur und –druck. Dies erfordert die Vorausberechnung der Druck- und Temperaturprofile entlang des Bohrlochs zur Identifizierung von möglichen Phasenübergängen/-wechseln und Dichtesprüngen, die Bestimmung von Bohrlochdruck und –temperatur häufig erschweren.

In der Erdölindustrie wird die Bohrlochströmungsmodellierung und –simulation häufig genutzt um die Druck- und Temperaturprofile entlang/längs des Bohrlochs vorauszuberechnen.

Die existierenden Quellcodebibliotheken/Programme wurden für grundlegende Lagerstättenfluide entwickelt.

Für diese Fluide bestand bisher nicht die Notwendigkeit zu einer erhöhten Genauigkeit bei der Vorausberechnung des Temperaturverlaufes entlang des Bohrlochs. Da es nur eine geringe Abhängigkeit des Druckes von der Temperatur gibt, ist dies (vereinfacht) gerechtfertigt.

Desweiteren verbleiben diese Fluide im Allgemeinen in der flüssigen Phase (Wasser und Öl) oder in der überkritischen Phase. Bei Kohlendioxid mit einer kritischen Temperatur von 30,04°C ist dies nicht der Fall, weswegen es vermutlich zu einem Phasenübergang entlang des Bohrlochverlaufes kommt.

Phasenübergang macht die Abschätzung von Temperatur und Druck schwierig und sorgt für eine Beeinträchtigung der Strömung. Zur Sicherstellung einer stabilen und leicht vorhersagbaren Strömung ist es üblich Kohlendioxid in überkritischem Zustand zu injizieren.

Bei einigen kostenintensiven Maßnahmen werden die oberen Teile der Injektionsbohrung gedämmt, um ein Abkühlen zu verringern, was wiederum zu Phasenübergängen führen würde.

Eine Folge der nicht konsequenten Abschätzung der Fluidtemperatur entlang des Bohrlochs ist die Vernachlässigung der Abhängigkeit der Fluidtemperatur von der Injektionsdauer.Mit zunehmender Injektionsdauer erhöht sich der Temperaturausgleich zwischen Fluid im Bohrloch und dem umgebenden Gestein.

Dies beeinflusst das Temperaturprofil des Bohrlochfluids und führt unter Umständen zu einer deutlichen Änderung im Dichteprofil und damit im Druckprofil.

Ein weiterer Grund für die verringerte Genauigkeit mit der die bisher gebräuchlichen Routinen/Programme den Druck und die Temperatur entlang des Bohrlochs berechnen, liegt in der Verwendung von klassischen/üblichen, räumlichen Zustandsgleichungen und Korrelationen zur Berechnung der thermodynamischen bzw. der Transport-/Bewegungseigenschaften.

Die ungenügende Genauigkeit dieser verwendeten Gleichungen und Korrelationen insbesondere in der flüssigen Phase und im Bereich des überkritischen Zustands sind bekannt. Folglich sind die Berechnungen der thermodynamischen und der Transport Eigenschaften mit großen inhärenten Fehlern behaftet.Desweiteren werden Druck und Temperatur mit numerischen Verfahren berechnet.Diese Approximationen erhöhen die möglichen Fehler bei den berechneten Werten von Druck und Temperatur.

Die aufgeführten Punkte rechtfertigen die Notwendigkeit für neuartige und belastbare Modelle und Lösungen, die nicht nur präzise die Temperatur und den Druck entlang des Bohrlochs als Funktion von der Bohrlochtiefe und Zeit, als auch Phasenwechsel und deren Einfluss auf die Temperatur und den Druck entlang des Bohrlochs berechnen können. Die Ziele der Arbeit sind geeignete Modelle und analytische Lösungen zur Berechnung von Druck und Temperatur im Verlauf des Bohrlochs. Im Zuge der vorliegenden Arbeit wurden ein Modell zur Berechnung des Drucks und vier Modelle zur Berechnung der Temperatur entwickelt.

Das zentrale/grundlegendste Modell zur Berechnung der Temperatur des Bohrlochverlaufes ist das stringenteste/konsequenteste Modell. Es ist nicht adiabatisch und als solches berücksichtigt es die Temperatureffekte durch Kompression/Ausdehnung des Fluids im Bohrloch und die durch den Temperaturausgleich zwischen Fluid und dem umgebenden Gestein. Für das Bohrloch und das umgebende Gestein wird die Annahme eines Wärmeübertragers als eine Verbundwand zwischen

Wärmequelle und –senke angenommen. Die Verbundwand besteht aus Bohrlochkomponenten (üblicherweise der Zementierung, der Verrohrung, dem Ringraum und dem Produktionsrohr).

Die drei weiteren Temperaturmodelle sind das adiabatische, das geothermische und das isothermische Modell.

Beim adiabatischen Modell wird die vollständige Isolierung vom umgebenden Gestein angenommen. In folgedessen wird die Temperatur nur durch Kompression und oder Ausdehnung des Fluids im Bohrloch beeinflusst.

Beim geothermischen Modell wird angenommen, dass die Temperatur des Fluids, der des umgebenden Gesteins der jeweiligen Teufe entspricht. Dies ist der Fall bei sehr geringer Fließrate und entsprechend langen Kontaktzeiten zwischen Fluid und Rohroberfläche.

Das isothermische Modell entspricht einer gleichbleibenden Temperatur entlang des Bohrlochs. Dies ist bei einer horizontalen Bohrung gegeben, bei der die Einlauftemperatur der geothermischen Temperatur entspricht und die Druckverluste in Folge von Reibung so klein sind, so dass es zu keiner bedeutenden Temperaturerniedrigung kommt.

Die Verbindung des nicht adiabatischen Temperatur-Modells mit dem Druck-Modell bildet die Basis für die konsequente/verbesserte Druck- und Temperatur-Berechnung bei einer Gasinjektion beliebiger Ausstattung und Ablenkung und im Falle von verschiedenen geologischen Formationen mit unterschiedlichen thermischen Leitfähigkeiten. Neben der analytischen Berechnung von Druck und Temperatur, wird die hohe Genauigkeit durch die Implementierung von hochpräzisen grundlegenden Zustandsgleichungen der thermodynamischen und der Transport Eigenschaften von Kohlendioxid gewährleistet. Dies sind die Gleichungen von Span und Wagner, von Fenghour et al. und von Scalabrin et al.. Diese werden genutzt um die thermodynamischen Eigenschaften, die Viskosität und die Wärmeleitfähigkeit zu berechnen.

Die Beschränkungen der Gleichungen von gesättigtem Kohlendioxid bedeuten, d aß sich diese Arbeit nur auf die Verwendung von Kohlendioxid bezieht. Durch die Verwendung der adäquaten Zustandsgleichungen anderer Gase bzw. Gasgemische lässt sich die Arbeit auf diese anwenden/um diese erweitern.

Um die Transparenz zu erhöhen und eine bessere Vergleichbarkeit sicher zu stellen, wurden dimensionslose Zahlen verwendet. Diese Zahlen vereinfachen die Analyse der Ergebnisse und nutzerverursachten Fehlern.

Die wichtigsten Neuerungen sind im Folgenden zusammengefasst:

1. Entwicklung eines Arbeitsablaufes zur Nutzung von hochpräzisen multiparametrigen Zustandsgleichungen zur Berechnung der thermodynamischen und der Transport Eigenschaften entlang des Bohrlochs.
2. Überarbeitung der bestehenden Arbeitsabläufe zur Berechnung des gesamten Wärmeübertragungskoeffizienten zwischen Gestein und Fluid.
3. Anpassung eines neuen analytischen Modells zur Temperaturabsenkung beim Bohrlochtest zur Vorhersage der Temperaturverteilung im Gestein zu jedem Injektionszeitpunkt.
4. Überführung der Druck- und Temperaturgleichungen in Form von dimensionslosen Zahlen zur Vereinfachung und besseren Transparenz der Lösungen, leichterer Leistungskontrolle der Programme, Lösungsanalyse und zur Identifizierung von nutzerverursachten Fehlern.
5. Überarbeitung der numerischen Algorithmen und die Implementierung von analytischen Berechnungsverfahren für eine verbesserte Vorausberechnung von Druck und Temperatur entlang des Bohrlochs.

Die Lösung der Modelle wurde durch die Simulations Programme implementiert, die mit Hilfe der Visual BASIC.NET Programmiersprache auf der Microsoft Visual Studio Entwicklungsumgebung. Die Programme sind in einem anwenderfreundlichen grafischen Nutzerinterface eingebettet. Diese Schnittstelle gewährleistet eine einfache Datenein- und –ausgabe.

Um gleichzeitig eine hohe Rechengeschwindigkeit bei guter Genauigkeit zu erreichen, wurden Tabellen der thermodynamischen und der Transporteigenschaften für die relevanten Temperatur- und Druckbereiche für eine CO_2 Injektion in geologische Formationen erstellt(Temperatur 250K-450K und Druck 1bar-400bar). Bei dem maßgeblichen Druck und Temperatur im Bohrloch werden die Eigenschaftsdaten vom Programm eingelesen und nach Bedarf interpoliert.

Der „marching" Algorithmus wurde implementiert, um Fehler im Zusammenhang mit Durchschnittswerten der thermodynamischen und der Transporteigenschaften für analytische Lösungen von Bohrlochdruck und –temperatur zu minimieren. Dies beinhaltet die Segmentierung des Bohrlochs, bei welchem die Längen determiniert werden durch ein Validierungskriterium, welches die hohe Genauigkeit der Lösung sicherstellt. Das Kriterium erfordert eine relative Differenz von maximal 0,1% bei dem z-Faktor zwischen Ein- und Ausgang des jeweiligen Segments. In der Folge der Segmentierung des Bohrlochs wurde eine Programmschleife programmiert, die Rechenschritte vom Ein- zum Ausgang des Rohres wiederholt.

Die Vorausberechnung durch die Routinen sind mit den gemessenen Daten von zwei Kohlendioxidinjektionsprojekten in salinaren Aquiferen unter verschiedenen Injektionsbedingungen verglichen worden. Bei dem ersten Projekt handelt es sich um das „onshore –Projekt Ketzin“ im Norden Deutschlands. Acht Injektionstage mit annähernd ähnlichen Bohrlochbedingungen (Temperatur, Druck und Fliessrate) wurden vom Datensatz ausgewählt. CO_2 wurde mit geringen Raten (ca. 30.000 m3/Tag) durch eine vertikale Bohrung in die in 630 m Teufe liegende Lagerstätte injiziert. Das Fließen erfolgt vorwiegend in der gasförmigen oder in gasähnlichen überkritischen Phasen.

Das zweite Projekt ist das Snøhvit -Projekt in Norwegen. Bei diesem Projekt wurde CO_2 durch eine abgelenkte „offshore“ Bohrung mit ca. 2600m tatsächlicher vertikaler Teufe injiziert. Elf Tage der Injektion mit ähnlichen Bohlochbedingungen wurden aus dem Datensatz ausgewählt. Das CO_2 floss mit einer verhältnismäßig hohen Rate (ca. 500.000 m3/Tag) in der flüssigen und flüssig ähnlichen überkritischen Phasen.

Für diese zwei Fälle liegt die relative Abweichung von simulierten und gemessenen Werten zwischen 0,01% und 1,93% bei den Drücken und für die Temperaturen zwischen 0,33% und 10,83%.

Thermische Leitfähigkeiten des umgebenden Gesteins waren die hauptsächlichen übereinstimmenden Parameter für die Temperatur im Bohrloch. Die beste Übereinstimmung für die Bohrung in Ketzin wurden erreicht durch die Berücksichtigung, dass das umgebende Gestein drei verschiedene Schichten mit unterschiedlichen Wärmeleitfähigkeiten hat, insbesondere 7W/mK, 6W/mK und 2W/mK. Die hohen Wärmeleitfähigkeiten sind im oberen Teil des Bohrlochs und deuten auf harte oder hoch verdichtete/verfestigte und sehr gering porösen Sandstein hin. Die beste Übereinstimmung wurde bei Snoehvit, unter der Annahme der Äquivalenz der Wärmeleitfähigkeit von 1,25W/mK der Gesteinsschicht, erreicht. Die geringe Wärmeleitfähigkeit der Felsformation stimmt mit der Tatsache überein, dass die Snoevhit Bohrung unter Wasser ist und von verhältnismäßig stark verfestigtem Sand mit hoher Porosität und hohem Wasseranteil umgeben sein dürfte.

Der Übereinstimmungsprozess deutet auf eine starke Abhängigkeit der Fluidtemperaturen und des umgebenden Gesteins zu den Wärmeleitfähigkeiten des Gesteins. Dies wird in dieser Arbeit bestätigt durch die Sensitivitätsanalyse von Fluid und Gesteinstemperaturen in Relation zur Wärmeleitfähigkeit. Es bestätigt sich die Tatsache, dass die Arbeit genutzt werden kann um die Wärmeleitfähigkeit des Gesteins abgeschätzt werden kann, immer wenn Temperaturmesssungen des Fluids im Bohrloch verfügbar sind.

Die weitere Untersuchung der Genauigkeit der Arbeit erfolgte durch den Vergleich der Vorhersagen mit den Messungen einer CO_2-produktionsbohrung. Die Bohrung hat einen langen Abschnitt mit einer 26,5% Ablenkung.

Die Ergebnisse sind ebenfalls sehr präzise.

Die maximalen Abweichungen für den Druck und die Temperatur sind 3 und 4%. Dennoch ist die Feststellung wichtig, dass eine allgemeine Implementierung der Programmierung bei Produktionsbohrungen, eine Berücksichtigung der Phasenverschiebung im Fall von hohem Flüssigkeitsanteil berücksichtigen sollte. Eine Phasenverschiebung wurde bei dieser Arbeit nicht berücksichtigt.

Einen vertieften Einblick in die Übereinstimmung von Modell und Theorie wird durch den Vergleich der Vorhersage/Vorausberechnung des nicht adiabtischen Temperatur-Modells mit den zuvor erwähnten Temperaturmodellen erhalten.

Es lässt sich schlussfolgern, dass sich diese Arbeit mit den theoretischen Prinzipien der Strömungsmechanik in zylindrischen Rohren und der Wärmeübertragung in Wärmeübertragern (-tauschern) mit gemischten Außenwänden und in Festkörpern in Einklang befindet. Eine weitere Bestätigung der Theorie erfährt die Arbeit durch die Analyse der Vorhersagegenauigkeit der Injektionsparameter (Bohrlochtemperatur, -druck, Fließ rate, innerem Rohrdurchmesser und Oberflächenrauhigkeit) und Injektionszeit.

Mit dem Ausloben der Genauigkeit des in dieser Arbeit erarbeiteten Ablaufes der Vorausberechnung von Temperatur und Druck von CO_2 in einer Injektionsbohrung, ist es ebenfalls die Erkenntnis wichtig, dass in mehrfacher Hinsicht weitere Möglichkeiten zur Verbesserung existieren.

Die Verbessserungen sind eng an die untermauernden Annahmen der Temperatur- und Druckmodelle gebunden.

Eine dieser Annahmen ist sehr konservativ und vernachlässigt die vertikale Temperaturverteilung/-ausbreitung des umgebenden Gesteins und im Fluid. Eine weitere, mit einem empfindlichen Einfluss auf die Vorausberechnungen, sind die Verdampfungsenergie und die frei werdende Kondensationsenergie im Fall des Phasenübergangs und deren Verteilung vom Bohrloch in die Formation über die gesamte Länge des Bohrlochs. Dies ist nicht in vollem Umfang richtig, da ein Teil der thermischen Energie durch Konvektion übertragen wird. Desweiteren wird der Massentransport zwischen zwei Phasen-zwischen Gas- und Flüssigkeitsphase nicht berücksichtigt. Konsequenterweise sind die Modelle nicht in der Lage das Temperatur- und Druckprofil bei einer geschlossenen Produktionsbohrung siedend oder kondensierend, in akzeptabler Genauigkeit darzustellen. Dies ist deshalb von Bedeutung, weil shut in´s (Produktionsstopps) bei realen Injektionen im Feld vorkommen.

Mit Blick auf die weitere Vorgehensweise empfiehlt sich die Modellierung der Massen- und Wärmetransportphänomene in Analogie zu einer Destillierungskolonne. Wichtig wäre auch eine Betrachtung der vertikalen Temperaturverteilung im Gestein.

Weiterführende Arbeiten, die die Genauigkeit der Vorausberechnung signifikant erhöhen könnten, wären eine direkte Implementierung der multiparameter Zustandsgleichungen in Algorithmen mit höherer Effizienz. Die Routinen könnten für andere Gase oder Gasmischungen unter Beinhaltung von Tabellen mit den Eigenschaften, erstellt mit den zugeordneten multiparameter Zustandsgleichungen oder durch die direkte Umsetzung in Routinen.

Mit dem Phasenwechsel durch z.B. Strömungsverschiebungsmodelle würde es ermöglichen die Routinen zur Vorausberechnung von Temperatur und Druck in Gasproduktionsbohrungen mit hohem Flüssigkeitsanteil zu nutzen.

MIt Blick auf die Zukunft, wird festgestellt, dass, obwohl noch weitere Möglichkeiten für Verbesserungen gegeben sind, es sich gezeigt hat, dass die Vorausberechnung von Druck und Temperatur von CO_2 bei einer Injektionsbohrung als Funktion von Tiefe und Zeit verlässlich sind. Es hat sich auch eine genaue Vorausberechnung von Temperatur und Druck in einer CO_2 Produktionsbohrung mit geringem Flüssigkeitsanteil unter Vernachlässigung von Phasenänderungen bestätigt.

ACKNOWLEDGEMENTS

First of all I am very grateful to Prof. Günter Pusch who accepted me as a Doctor candidate and employed me as scientist in his research team some months before his retirement.

Next I am grateful to Prof. Leonhard Ganzer who after taking over from Prof. Push as head of department of reservoir engineering kept me in his research team and accepted to continue the supervision of my dissertation. He also incessantly provided support in many forms whenever he could.

I likewise feel obliged to Prof. Jan-Rune Ursin of the University of Stavanger in Norway, who happily agreed to collaborate with us as well as invited and assisted me throughout my studies in that university. He also strongly contributed to make my stay in Stavanger a pleasant one to remember. His inputs were very valuable to implement the analytical solutions to the pressure and temperature equations.

Furthermore I would like to express my special thanks to Dr. Jan Henninges of the GFZ in Potsdam and Statoil in Norway for making field measured data available to validate the models.

I would like to thank Prof. Wolfgang Wagner of Bochum for his advice on equations of state and the book on the GERG's wide range multi-parameters equation of state that he graciously offered me. The information in this book has been of very valuable help.

My great appreciation is also expressed to all my colleagues of the reservoir engineering department, Clausthal and EFZN, Goslar, especially to Mr. Kilian N. Awemo for the continuous exchange and challenge of ideas through brainstorming and discussions. I would also acknowledge the great help of Mr. Ralf Peix of EFZN in translating the summary of this work into German language.

I am also greatly thankful to the African community of TU-Clausthal and especially the Cameroonian community for their incessant supports in extra-curricular issues and activities. They have made living in Clausthal and Goslar bearable.

Further special thanks go to my friends who shared my happiness, my frustrations and sorrows as well as to my family who gave me the necessary support in the pursuance of higher education, especially during the tedious doctorate days.

I could extend this list a lot more but will stop here to give the reader the opportunity to read the work. I apologize for not having mentioned everyone with name who helped me in this successful endeavor. Many thanks to everybody

DEDICATION

This work is first of all dedicated to my late wife Cecile, who died in 2010 some months before the defense of her own PhD thesis. May her soul rest in perfect peace

It is also desiccated to my late father who could not live to see me and my siblings growing to become men and women. May his soul also rest in perfect peace

Table of Contents

LIST OF FIGURES

LIST OF TABLES

NOMENCLATURE

A	Cross-sectional area of the flow conduit
A_s	Surface area available for flow and for heat transfer
C_p	Heat Capacity at constant pressure
C_V	Heat capacity at constant volume
D	Diameter of the flow conduit
d	When preceding any variable it means the derivative of that variable
dev	Percentage deviation
E	Emissivity
E_i	Exponential integral
e	Pipe roughness
F	Force
F_{tci}	View factor between tubing outer surface and casing inner surface
$\vec{F}$	Force vector
f	Fugacity
f_f	Friction factor
$f(t)$	Transient function for heat propagation in the rock
G	Gibb's free energy

g Acceleration due to gravity

H Enthalpy

h_{an} Heat transfer coefficient by natural convection in the annulus

h_f Fluid film transfer coefficient by forced convection in the flow conduit

h_r Heat transfer coefficient by radiation in the annulus

I_{tci} Overall heat interchange factor between tubing outer surface and casing inner surface

J_H J-factor for heat transfer

K_f Formation thermal conductivity

K Thermal conductivity

L Length of a wellbore segment

l Length of the flow conduit

M Molecular weight

$\dot{m}$ Mass flow rate

N_f Friction number

N_g Gravity number

N_{Gr} Grashof number

N_{Nu} Nusselt number

N_{pr} Prandtl number

N_{Re} Reynolds number

N_{st} Stanton number

P Fluid pressure

Q Heat transferred per unit length in or out of the system

$\dot{Q}$ Rate of heat transferred per unit length

q	Volumetric flow rate
$\dot{q}$	Rate of molar heat transferred per unit length
R_{in}	Dimensionless radius of thermal influence
r	Radius
r_{in}	Radius of thermal influence
S	Entropy
T	Temperature
T^*	Absolute temperature for calculating irradiative heat transfer coefficient
$\overline{T}_f$	Mean fluid temperature
t	Time
U	Internal Energy
U_{over}	Overall heat transfer coefficient
u	Fluid velocity
V	Volume
W	Work done on the surrounding by the system or vice-versa
y	Well depth
z	Z-factor

Greek symbols

α	Well inclination angle in degrees
α_R	Rock thermal diffusivity
α_v	Volumetric thermal expansion coefficient
χ	Fraction of vapour in the fluid mixture
β	Thermal expansion coefficient
Δ	When preceding any variable it means the gradient of that variable

ε	Relative roughness of flow conduit
φ	Dimensionless number controlling temperature change as a function of hydrostatic pressure
η	Joule-Thompson coefficient
λ	Chemical potential
ϕ	Fugacity coefficient
ρ	Density
σ	Stefan-Boltzmann Constant
μ	Viscosity
π	Constant ~ 3.14
τ	Shear stress
ω	Number controlling temperature change as a function of frictional pressure

Subscripts

an	Annular space
av	Average
cas	Casing
bh	Bottom hole
cem	Cement
$crit$	Critical
F	Undisturbed formation
f	Fluid
fc	Formation-cement interface
fF	Fluid & formation
g	Gas
IG	Ideal gas

i	Index representing an integer or inner part
j	Index representing an integer
in	Inlet
L	Liquid
m	Mixture
o	Outer
out	Outlet
ov	Overall
$over$	Overall
R	Rock
r	Radius
s	Surface
sci	Casing inner surface
$scem$	Cement ring inner surface
std	Standard
sti	Tubing inner surface
tub	Tubing
w	Wellbore
wh	Wellhead

1. INTRODUCTION

1.1 Background Information

The eminence for reducing greenhouse gas emissions from the environment urged societies to devise means towards CO_2 capture and storage. Amongst such means geological storage is seen so far as one of the major short to medium-term alternatives (Bielinski, 2007, Rebscher and Oldenburg, 2005, Pruess, 2005, Benson and Cook, 2005).

The opportunities found in enhancing recovery of oil in oil reservoirs have led to intensive research on CO_2 properties that directly or indirectly affect injection process in general and flow in wellbore in particular. In order to get CO_2 in the reservoir being it for storage or for enhancement of hydrocarbons recovery, the fluid must transit through the well from the surface. In this respect, it is important to understand the behavior of the fluid on its way from the wellhead to the bottom of the borehole. This requires thorough investigation, which ideally should involve experimental and/or field measurements of most important parameters affecting the fluid behavior. In the absence of such measurements, modeling has proved over the past to be a powerful tool to estimate within experimental uncertainty, parameter which would otherwise be cost prohibitive to measure.

Models can roughly be grouped into two categories with respect to the approaches to solution: analytical and numerical models. Numerical models are complex. They incorporate as many details as possible. Therefore the number of underlying assumptions would be reduced. Conversely, analytical models are simpler. As a consequence, more assumptions would be considered in their development.

The solution to numerical models is obtained by approximation methods such as finite difference, finite volume or finite element. Such approximations would incorporate errors in the final solution. The magnitude of these errors usually depends on the method, the step sizes, the algorithm type and the size of the problem.

The analytical methods get the exact solution but are seldom appropriate for complex problems. The problem usually needs to be simplified with realistic assumptions for the solution to be found. An important issue is that sometimes, assumptions may over-simplify the problem and fail to capture its most important facets.

The development of analytical models to rigorously predict the temperatures and pressures of CO_2 along an injection well forms the subject of this dissertation.

The ensuing paragraphs present the detailed objectives and the scope of this work. They are followed by the description of the methodology used in attaining these objectives. Subsequent sections provide a review of the literature on fluid-flow in a wellbore in general and CO_2 flow in particular. The later sections firstly present the model development process. It is followed by the model validation part. In this part the predictions of the models are tested against field measured data and sensitivities to injection conditions and injection time are performed. The work is ended with general conclusions that wraps up the findings and with some recommendations for further works.

1.2 Objectives and Scope

Brill and Mukherjee (1999) made and extensive review of fluid flow in wells. They reported that many models have been proposed for single and multiphase flow prediction. Only few of them focus on gas injection in general and on CO_2 injection in particular. Conversely, the literature is very verse with empirical and mechanistic models for two and single phase upward flow of petroleum fluids in wells. The empirical and mechanistic models require modeling of flow regimes to capture multiphase flow behavior. The approaches to solution consider a one-dimensional multiphase flow problem and are generally based on fully implicit numerical schemes associated with well segmentation. Holmes et al (1998) recognized that these approaches are appropriate only when the functions are continuous and differentiable over the entire range of variables. They went on to propose two modeling approaches that overcome this discontinuity problem. These are:

1. The homogenous model: it assumes all phases to flow with the same velocity. Consequently, the properties of the fluid are estimated as weighted average of the values of each property from all the phases.
2. The Drift flux model: it takes into consideration the difference in velocity amongst the phases. As such, it provides a means to account for the slip between these phases.

Most of the existing codes for prediction of flow in wells have been developed with respect to petroleum fluids. As such they do not consider phase transition along the pipe nor do they model evolution of fluid's temperature and that of the surrounding rock with time. Paterson et al (2008) recognized phase transition as an important issue for CO_2 flow in geological formations. This issue becomes more re-sounding in offshore applications, where external temperature gradient in the sea is far different from the atmospheric and the geothermal gradients. Furthermore, these codes solve pressure and enthalpy balance equations with numerical methods. These methods would most often add errors to the results.

Some authors have proposed analytical solutions to pressure and enthalpy balance equations along the well. Amongst these, the solution of Hagoort (2005) is worth mentioning. Works have also been done in other branches of engineering and in other aspects of mining and petroleum engineering, on analytical solutions to problems in heat exchange and temperature propagation in composite solids and rocks. These solutions could be adapted to predict the evolution of temperature of fluid in the well and that of the surrounding rock as a function of injection days. Worthy to mention are the works of Eppelbaum and Kutasov (2006) on temperature drawdown well testing and those reported by Coulson and Richardson (1996) on heat transfer through composite walls.

In the light of the forgone points, the major objective of this work is to develop a rigorous procedure that is based on analytical solutions to pressure and enthalpy balance equations in order to accurately predict pressure and temperature of CO_2 in a well during injection into a geological formation as a function of depth and time. In achieving this main objective the following sub-objectives are also to be achieved:

1. Modeling phase transition and its associated effects on temperature and pressure in the well.
2. Revisiting existing analytical solutions to problems in heat exchange and temperature propagation in composite walls and rocks and adapting them to model the fluid temperature in the well and the propagation of temperature in the surrounding rock with time and with depth.
3. Improving the prediction of thermodynamic and transport properties of CO_2 in the well
4. Developing a simulation tool that can accommodate wellbores of varying diameters, varying deviation angle and non-uniform tubing string.
5. Developing a simulation tool that can be used to estimate thermal conductivity, thermal diffusivity of the surrounding rock at specified depths whenever temperature measurements are available. As such the codes can accommodate layered formations with different thermal properties and various geothermal gradients.

The model focuses on pure CO_2. Therefore a single phase solution is essentially adopted. However two-phase flow is captured through the homogenous fluid model. This neglects the effect of slip between flowing phases by considering fluids' properties to be the weighted average of the properties from the two phases.

1.3 Methodology

Hagoort (2005) proposed an analytical solution to temperature in a gas production well. He recognized that this solution if combined with an appropriate solution to pressure could form the building block for a more rigorous general prediction of temperature and pressure in gas production well. Recognizing the pertinence of this point made by Hagoort and based on the evidence he brought in validating his model, this work has followed his recommendations and some of the steps used in developing his model. As such analytical solutions to temperature and pressure have been developed and combined with accurate methods to estimate thermodynamic and transport properties to form a procedure to rigorously predict CO_2 pressure and temperature in an injection well.

Hagoort further rearranged the pressure and temperature equations to isolate dimensionless numbers in order to improve the transparency and control of his solutions. The same approach has been used in this work. Analogous dimensionless numbers have been isolated with respect to injection situation. They do not only improve the transparency and control of the solutions but they also make it easier to analyze the results and to detect user-related errors. This forms one of the innovations in this work.

Four solutions to temperature and one solution to pressure have been developed. The main temperature solution is the most rigorous in estimating the temperature along the well. It is non-adiabatic and as such it takes into account the temperature effects caused by compression/expansion of the fluid in the well and those related to the heat exchanged between the fluid and the surrounding rock. In this way the wellbore and the rock are assumed to form a heat exchanger with composite wall between the heat source and the heat sink. The composite wall is made up of the wellbore components (cement, casing, annulus and tubing).The three other temperature models are the adiabatic model, the geothermal model and the isothermal model.

The adiabatic model assumes that the well is perfectly insulated from the surrounding rock. Consequently only the effect of compression and/or expansion influences the temperature of the fluid in well.

The geothermal model assumes that the temperature of the fluid is equal to that of the surrounding rock at each depth. This situation would be approached whenever the flow rate is very low to allow longer contact time between the fluid and tubing surface.

The isothermal model assumes that the temperature is constant along the well. It would be approached in a horizontal well when inlet temperature is equal to geothermal temperature and frictional pressures losses are sufficiently small not to cause significant temperature drop.

In arriving at the temperature solutions and at the pressure solution some other innovative approaches were necessary not only to improve existing models but also to enable using them in the solution procedure. The major innovations that significantly contributed to the rigorous solutions in this work are the following:

1. Development of a workflow to use high accuracy multiparameters equations of state to predict CO_2 thermodynamic and transport properties along the well. This required creating runtime tables from which the properties are read during simulation.

2. Revisiting Willhite's procedure to calculate the overall heat transfer coefficient (OHTC) in the wellbore (Willhite, 1967). In this procedure the OHTC is expressed as a function of the outer surface area of a characteristic area which could be the outer area of the tubing (for injection through tubing) or inner area of the casing (for injection through annulus). This choice neglects the influence of the surface areas of other components and might result in a serious underestimation of the heat transfer area. This work capitalizes on the analogy between the wellbore and a heat exchanger with composite wall material discussed by Coulson and Richardson (1996) to estimate the OHTC as a function of an overall heat transfer area. The overall heat transfer area is calculated as the log average of the surface areas of the wellbore components from the tubing to the interface between the rock and the cement.

3. Adapting an innovative model for temperature drawdown well testing to predict temperature propagation in rock at any injection time being it early or late. This has been possible by assuming the flowing fluid to be an infinite cylindrical heat source or heat sink and by making analogy between the thermal contact resistance and the overall heat transfer coefficient. This is very important because it enables to estimate thermal conductivity and thermal diffusivity of the surrounding rock based on the measured temperature of the fluid in the well.

4. Revisiting the marching algorithm proposed by Brill and Mukherjee (1999 , P. 26) and implementing with analytical solutions to arrive at accurate estimates of pressure and temperature along the well. This involved including in the original algorithm the validation criterion proposed by Economides et al (1993, P. 145) for pressure in a gas well. This work also uses this criterion to validate temperature. The criterion requires that the relative difference between the z-factor at the inlet of a segment and that at its outlet should not exceed 0.1%.

The equations for pressure and temperature calculation are coupled and solved simultaneously to arrive at the solutions for pressure and temperature along the well. The pressure calculation is based on the extended Bernoulli's equation. It accounts for frictional pressure losses and neglects kinetic energy effects. The temperature of the fluid along the well is estimated with the enthalpy balance equation. It considers heat transfer between the fluid and the surrounding as well as the heating or cooling resulting from phase change and the heating as result of compression. The heat exchange with the surrounding follows the approach proposed by Ramey (1962) and revisited by Hagoort (2004). It assumes steady state heat flow through the wellbore components (cement, casing annulus and tubing) and transient heat flow through the formation. The heat flows through the wellbore components as a function of the overall heat transfer coefficient and the overall heat transfer area. Its propagation in the surrounding rock is controlled by a time function based on the thermal diffusivity equation in rocks.

Because of the inadequacy of the Ramey's time function to model heat propagation in the formation for times less than one week, this work has adapted the models proposed by Eppelbaum and Kutasov (2006)'s in analyzing temperature drawdown during well testing to derive a time function applicable to all times (earlier or later than one week). It assumes the fluid in the well to be an infinite cylindrical heater and draws on the analogy between the thermal contact resistance and the overall heat transfer coefficient.

Recognizing that the capability of a code to predict pressure and temperature of a fluid in a wellbore also depends on its ability to properly model thermodynamic and transport properties of the fluid, these properties are estimated with reference equations of state for CO_2 in order to obtain improved accuracy. The Span and Wagner (1996) equation is used for the calculation of density and other thermodynamic properties. The equations of Fenghour et al. (1998) and Scalabrin et al. (2006) are used for the calculation of viscosity and thermal conductivity respectively. These equations are computationally time-intensive. As such to avoid implementing them directly in the codes they are used to generate runtime tables from which the properties are read during simulation. The tables are generated in temperature and pressure ranges of 250 to 450 K and atmospheric pressure to 400 bar respectively. The simulator reads prevailing temperature and pressure in the wellbore and extracts the corresponding properties from the tables. Linear interpolation is used for values that fall in between those tabulated.

The solution to the models is implemented through simulation codes developed with the Visual Basic.Net programming language on the Microsoft Visual Studio platform. The codes are encapsulated in a user-friendly graphical user interface. This interface ensures easy input and output of the data.

The implementation of the marching algorithm enables to minimize the errors related to the use of average values of thermodynamic and transport properties in analytical solutions to pressure and temperature in the well. The algorithm consists of dividing the wellbore into segments of which lengths are determined by the above-mentioned validation criterion on temperature and pressure to ensure high accuracy in pressure and temperature solutions. Whenever the criterion is violated, the segment's length is chopped and calculation is repeated so on and so forth until the criterion is obeyed. A consequence of the segmentation of the well is the establishment of a computational loop which is repeated from the inlet to the outlet of the pipe.

The validation of the models and the procedure has been performed firstly by comparing the predictions of temperature and pressure with measured values from two injection projects in saline aquifers. Secondly same comparison has been made with measured data from a CO_2 production well to confirm the robustness of the solutions. Thereafter a comparison of the predictions of the main temperature model (non-adiabatic) to those of the three other models (geothermal, adiabatic and isothermal) is performed. And finally the sensitivities of the pressure and temperature predictions to injection parameters (wellhead temperature, wellhead pressure, injection flow rate and tubing diameter and inner roughness) and injection time is also analyzes. The aim of the temperature models' comparison and that of the sensitivity analysis is to extensively test the procedure for the predictive capabilities and agreements with general theories of fluid flow in cylindrical pipes and heat transfer into a heat exchanger with a composite wall.

2. LITERATURE REVIEW

The literature is rich in models for predicting single and multiphase flow in wellbores. These models differ from their nature and their underpinning assumptions. Specific applications to CO_2 have not been extensively reported. Notwithstanding, many works have been documented on the estimation of the properties of CO_2 that play important roles on its state and behavior along a pipe in general and in a wellbore in particular. The following paragraphs will first review major works on fluid flow in pipes with an emphasis on the developments with respect to the estimation of pressure and temperature along the wellbore. Thereafter they will highlight important contributions towards the prediction of thermodynamic and transport properties of CO_2.

2.1 Wellbore Flow Modeling

Wellbore flow modeling has evolved over the years thanks to the efforts in predicting pressure and temperature profile of single and multiphase fluids in oil and gas production wells. Brill and Mukherjee (1999) made an intensive and extensive review of the theories underpinning multiphase flow in wellbores and pointed out most relevant works on the subject. Brill and Arirachakaran (1992) earlier presented the historical development of multiphase flow analysis to which they distinguished three main periods: the empirical period (1950-1975), the Awakening Years (1970-1985) and the nowadays modeling period, which started from 1980's.

Ayala et al (2007) discussing the Brill and Arirachakaran's classification pointed out that: "*during the empirical period, all efforts were focused on correlating data from laboratory and field facilities in an attempt to encompass the widest range of operational conditions as possible*". They also reported that an earlier attempt to empirically predict two-phase flow pressure drop in horizontal pipes is the work of Lockhart and Martinelli (1949). Their work was improved and extended to pipes of other inclinations by some other authors. One of the most significant extensions is that of Beggs and Brill (1973) who considered horizontal, vertical and inclined pipes in their works and also incorporated the Froude number, a dimensionless number that measures the influence of gravity on fluid motion. Brill and Mukherjee (1999) recognized that it is the advent of personal computers and the understanding of techniques to numerically integrate the pressure gradient along a pipe that improved the capabilities in the prediction of multiphase flow in the wellbore. The new capabilities that also marked the start of the modeling period, created the necessity to introduce more physics fundamentals in the multiphase flow analysis (Ansari et al., 1994, Kabir and Hasan, 1990). This period benefited from advanced multiphase flow technologies already existing in nuclear and chemical engineering. As a consequence, the period was marked by the emergence of two-phase transient and steady state mechanistic[1] models. These models are able to describe physical phenomena with improved accuracy. The accuracy of the models was further enhanced by the introduction of compositional fluid modeling. This was favored by increasing accessibility to computers as well as their increasing user friendliness, capacity and speed. Brill and Mukherjee (1999) further anticipated that another technological development that would significantly contribute to advances in multiphase flow modeling is the progress in data acquisition and measurement technology. The progresses in technology have been more significant in the recent past and computer speed has tremendously increased to permit analysis of flow in a more detailed manner.

[1] Models trying to capture specific features of individual flow patterns in which simplified conservation equations are solved, with the main focus being the prediction of pressure drop and hold-up (Ayala et al., 2007)

In this light Ayala et al (2007) identified that the state of the art of fluid flow modeling would be based on the principles of continuum mechanics and thermodynamics.

The prediction of temperature and pressure in gas wells is often based on solution of the extended Bernoulli equation and the steady state energy balance equations. The approach to solution is essentially grounded on one of the numerical integration methods that couple pressure gradient and enthalpy gradient equations. Brill and Mukherjee (1999) recognized that computer programs mainly use the "Marching Algorithm" to perform pressure traverse calculations. The algorithm starts at either top or bottom of the well and numerically integrates pressure and temperature gradients over successive small length sections, marching to the other end.

Ayala et al (2007) expatiated on multiphase flow. They pointed out that there are four major widely used multi-phase fluid models. These depend on the way the pressure and temperature are modeled and on how the interactions between phases are described.

1. The homogenous model: this model treats fluid as single phase with properties equal to the weighted average of the properties of the different phases in presence.
2. The drift-flux model: this model considers the average local velocity difference between the fluids. Therefore Holmes et al. (Holmes et al., 1998) also referred to it as homogenous model with slip.
3. The separated model: in this case, the phases are assumed to be flowing in separated zones of the pipe and move at different velocities in the same or opposite directions.
4. The Multi-fluid models: this model takes into consideration the phases in place and the interactions between them.

Two fluid modeling approaches are used to account for composition and thermodynamic and transport properties. These are the blackoil and the compositional approaches. Black oil modeling is the most widely implemented. It treats the material balance equations in term of fluid phases, whereas compositional modeling accounts for all the components and phases present.

Black-oil modeling assumes fluid components to be lumped into pseudo-components represented by a maximum of three phases (oil, water and gas), which are identified by their specific gravities. PVT properties are modeled as simple functions of pressure and temperature. As a consequence empirical correlations are often used to estimate density and viscosity. Overlooking detailed fluid composition raises the question of model validity whenever dealing with multi-component volatile fluids. For these fluids it is always necessary to consider the effect of composition on fluid behavior. A detailed compositional modeling is therefore required. Furthermore blackoil models also rely on empirical correlations in the determination of flow characteristics (patterns). Such correlations have been proposed by Duns (1963) , Hagedorn and Brown (1965), Orkiszewski (1967) and Beggs and Brill (1973) amongst other authors. Because of their empirical nature, the range of applicability of these correlations depends on several factors strongly related to the experimental conditions. Such parameters are tubing size, oil gravity, gas-liquid ratio and water-cut amongst others. This dependency precluded their generalization to include cases outside the range of experimental parameters.

With the compositional approach the variation of the fluid composition in each phase along the wellbore can be captured. This variation is a function of temperature, pressure and eventual slip between the phases. Therefore phase behavior and vapor-liquid equilibrium calculation play an important role. The identification of the flowing phases and their compositions enables to calculate the thermodynamic and transport properties of the fluid. These calculations are usually performed with equations of state and transport properties models.

Several authors amongst whom Gregory and Aziz (1978), Anis and Buthod (1974), Gould (1979) and Furukawa et al. (1986) presented early compositional modeling approaches to simulate fluid flow in pipelines. The application of compositional modeling to multiphase/multi-component fluid flow in wellbore has not been widely reported in the literature. Two limiting models can be applied to compute composition of a moving multi-component fluid in the wellbore (Pourafshary et al., 2008). The first one is the equilibrium model, which assumes equilibrium between vapor and liquid phases at all positions in a wellbore. This assumption makes it is possible to perform flash calculation in each segment of the well to obtain fluid composition. The second one is the differential vaporization model. It is based on the premises that there is no transfer of vapor components back into the liquid.

Pourafshary et al (2008) further propounded that two of the multi-phase fluid models earlier discussed by Ayala et al. (2007) are widely used to model phase fractions and pressure distribution in multiphase-flow modeling in wellbores. These are the homogenous fluid model and the separated fluid model. They went on to recognize that assumption of equilibrium between liquid and gas in the homogenous model is reasonable due to the consideration of no-slip behavior of the phases. Conversely, such assumption would be unreasonable for the separated models, which consider a slip velocity between phases. The gas phase moves faster than the liquid phase at each section rendering the equilibrium assumption doubtful. Anis and Buthod (1974) reported that the same issues play a role in fluid-flow modeling in pipelines. They showed that in a pipeline, assumption of equilibrium model always results in more pressure drop than differential model. They also studied the effect of these models on vapor velocity and concluded that the differential model predicts a lower velocity than the equilibrium model.

Pourafshary and his co-authors further identified that the actual flow condition in the wellbore falls somewhere between the two extremes of equilibrium and differential vaporization models. In this light, they proposed modifications for flash calculations to account for slip between the phases. To implement their modifications of flash calculation to the simultaneous modeling of phase slip and vapor-liquid equilibrium, they coupled wellbore flow simulator with a numerical reservoir simulator. They reported that this model accurately computes compositional fluid flow between the wellbore and reservoir.

Analytical solutions to wellbore flow equations have also been documented in the literature albeit not as extensively as the numerical solutions. These are mostly limited to single phase flow or treat the multiphase fluid as a homogenous fluid with local properties equal to the weighted average of properties of the phases in presence. In this respect Economides et al. (1993) proposed practical solutions of flow equations in wells. These equations take the average fluid properties over the entire well length. Therefore to minimize errors, they advise to segment the well into an appropriate number of segments required to achieve desired accuracy. Hagoort (2005) also proposed analytical models for prediction of pressure and temperature in a gas production well. His solution is strongly based on dimensional numbers and suggests a strong incorporation of the marching algorithm. He recommends that thermodynamic and transport properties at prevailing temperature and pressure be read from tables. These tables are generated with suitable equations of state and transport properties equations. Linear interpolation is often required when the temperature and pressure values fall in between the tabulated ones.

The dimensionless number approach proposed by Hagoort (2005) enables better control and understanding of the models. Therefore similar approach is used in this work to isolate parameters that control frictional and hydrostatic effect on pressure and temperature. In this way the models' performance can easily be inspected and the possible abnormalities could be readily identified. During injection the effect of slip between liquid and gas phase is not as important as during upward flow in a production well. This is because the gravity effect that drags the liquid downwards and the acceleration of the fluid particles all contribute to drive the fluid downward. This justifies the use of homogeneous model to account for multiphase flow in an injection well. Furthermore the thermodynamic tables also enhance code efficiency while preserving high accuracy of the estimates. This is especially true when the entries are generated by accurate equations of state that are monotonously continuous in the operations range of temperature and pressure. These conditions are obeyed by the equations of state in this work. As such it is reasonable to use tables in estimating thermodynamic and transport properties.

2.2 State of the Art in Modeling CO_2 Flow in Pipes

To the knowledge of the author of this dissertation, there is no extensive literature on CO_2 flow modeling in pipes in general and in wellbores in particular. However, some important contributions in this direction have been reported.

Cronshaw and Bolling (1982) proposed a numerical model for non-isothermal flow of carbon dioxide in production and injection wells. This represents one of earliest works that attempts to rigorously predict CO_2 temperature and pressure in a well. The model considered heat exchange with the surrounding and evolution of CO_2 temperature with time (see Figure 2.1). However it did not attempt to predict temperature propagation in the surrounding rock. The model was successfully validated on a well producing CO_2 with 3% impurities.

Hein (1985) presented rigorous and approximate methods for CO_2 pipelines analysis. He highlighted the fact that CO_2 flows most effectively at pressures higher than its critical pressure in the so called dense-phase region. Majority of CO_2 injection projects for EOR purposes operated in this region, which presents no complexity for modeling and as such the models would follow the same principles as those used for single phase dense fluids. However, for general purposes and especially when the flow takes place partially or entirely outside the supercritical region, Hein (1985) recognized the necessity for a predictive model to allow for variation of thermodynamic and transport properties in the flow lines. He further highlighted the high dependence of CO_2 enthalpy on temperature and consequently seized the opportunity to advise consideration of isenthalpic expansion/compression (joules-Thomson effect) in an effort to improve the prediction of CO_2 temperature and pressure along the flow line.

Lu and Connell (2008) took to Hein's advice in their works and followed the same approach as Cronshaw and Bolling (1982). They proposed a numerical procedure to evaluate the non-isothermal flow of carbon dioxide and its mixtures in injection wells. As such they considered the heat exchanged with the surrounding rock as shown in Figure 2.1. Their procedure also models two-phase flow and phase transition in the well. The fluid and thermodynamic properties, including the saturation pressure of the gas mixture are calculated with the Peng-Robinson I (PRI) cubic equation of state. They highlighted the potential complexity of carbon dioxide wellbore flow hydraulics for sequestration applications. They also recognized the important role it plays in the determination of the bottomhole pressure and hence the injection rate. The model did not also attempt to predict temperature propagation in the surrounding rock.

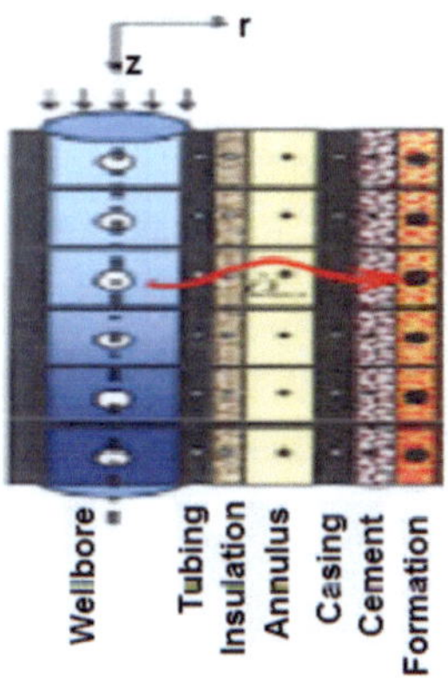

Figure 2.1*: Example of discretized wellbore system used in the numerical models of Cronshaw and Bolling and Lu and Connell (adapted from Bahonar et al., 2010)*

Paterson et al (2008) applied the model of Lu and Connell to calculate CO_2 density profiles for wells experiencing different thermal conditions to predict how bottomhole pressures are related to wellhead pressures. They identified that two phase or single phase fluid could be encountered in the wellbore. Therefore they re-emphasized on the necessity to account for evaporation/condensation and the associated heating and mass transfer effects. They also mentioned the importance of coupling wellbore and reservoir flow in order to effectively study the flow behavior of CO_2 along a well.

2.3 Thermophysical Properties of Fluids

Wakeham and Nieto De Castro (1996) give a clear definition of thermophysical properties. These are properties that will not involve the change of the molecular entities of the fluid during a process. Those include thermodynamic and transport properties. They suggest that thermodynamic properties relate to the differences between two states of thermodynamic equilibrium and as such they are those properties that determine the change in state of a fluid as a result of external stimuli. On the other hand, transport properties are those concerned with the rate of change of the state of fluid. This could be as a result of change in external conditions, or as a result of transport of mass, momentum or energy between different parts of a fluid in a non-uniform state. Table 2.1 below gives a classification of important thermodynamic and transport properties used in the context of this work flow modeling.

***Table 2.1**: Thermodynamic and transport properties used in the context of this work*

Thermodynamic properties		**Transport properties**	
Property	Stimulus	Property	Change trigger
Temperature	Heating/cooling	Viscosity	Heating/Cooling Compression/Expansion/shearing
Pressure	Expansion/compression	Thermal conductivity	Heating/Cooling
Fugacity	Expansion/compression		
Gibb's free energy	Heating/cooling, expansion/compression		
Chemical potential	Heating/cooling Expansion/Compression/elevation		
Molar Volume	Heating/cooling, expansion/compression		
Density	Heating/cooling, expansion/compression/elevation		
Enthalpy	Heating/cooling Expansion/Compression/elevation		
Entropy	Heating/cooling		
Joule-Thomson coefficient	Compression/expansion		
gas constant			
Specific heat capacity at constant Pressure			
Molar heat capacity	Heating/Cooling		

2.3.1 Thermodynamic Properties Models

Literature contains number of models to estimate thermodynamic properties of a gas. These are essentially the density (Z-factor as alternative), the specific heat capacities, the enthalpy, the entropy, the internal energy, the molar volume, the Joule-Thomson coefficient and the speed of sound. They are interrelated through the Maxwell relations[2] (Suzuki and Uematsu, 2002).They also all depend on the density which is the only independent property. As a consequence accurate estimation of the density is

[2] Maxwell relations are a set of thermodynamic equations that use the fact that the order of differentiation of a function of two variables does not matter, to state equality amongst second derivatives of the thermodynamic potentials.

the key to accurately predicting other properties. Two more reasons justify the need for accurate density estimation. Firstly, in a vertical or inclined well, the hydrostatic pressure is directly proportional to the density and secondly the viscosity and thermal conductivity are also functions of density.

Density prediction is essentially done with Equations of State (EOS). Many EOS have been reported in the literature. Fundamentally, two categories can be distinguished: Cubic EOS and Non-Cubic EOS. The Non-Cubic EOS's are based on more fundamental concepts and most often contain a very large number of parameters that make them to be referred to as multi-parameters equations of state (MPEOS). One of the most popular MPEOS for CO_2 is the equation of Span and Wagner (1996) which is based on Helmholtz free energy. This EOS covers the region from the triple-point temperature to 1100 K and pressures up to 800 MPa. Kim (2007) simplified this equation by reducing the number of its correlation coefficients. Notwithstanding the reduced number of coefficients, his equation also estimates CO_2 properties with high accuracy. Setzmann and Wagner (1991) proposed an EOS for pure Methane. Their equation is also based on Helmholtz free energy and covers a wide range of temperatures and pressures.

In general the MPEOS's involve extensive regression on experimental data to generate their multiple parameters. Therefore they are sometimes very complex and do not have simple and consistent mathematical forms. Most often, they also lack a unified functional form. As a consequence, their mixing rules are difficult to generate and therefore renders their extension to mixtures very complex. Mixing rules are specifically required for each newly introduced parameter. These points have been recognized by Frey et al (1991) who propounded that if an equation does not have a consistent mathematical form, it is difficult if not impossible to define appropriate mixing rules. Notwithstanding, Kunz et al.(2007, P. 43) proposed mixing models for the MPEOS. These models enabled them to develop multi-parameter equations of state for mixtures by combining the MPEOS of the individual components. Consequently these equations are limited to mixtures of components for which a MPEOS exists. They also result to the equation of a pure component in the extreme case, where the mixture consists only of the component in question. Kunz and his co-workers showed that their equations achieve accuracy that has never been obtained before for a natural gas mixture.

The complexity and difficulty to extend to all mixtures make these equations difficult and inefficient to implement in a simulation code. As a consequence they are still not widely accepted and applied. They are rather often used as benchmark to more efficient equations of state in absence on measured data. However their accuracy is of great advantage when dealing with a pure component or with mixtures for which all components have a MPEOS. This is especially relevant when pressure and temperature are anticipated to fall in the liquid and near critical regions.

The general practice, which is also the standard in the petroleum industry, is the use of semi-empirical formulae based on the corresponding states principle[3] to estimate PVT properties. The formulae have a simple mathematical cubic form that justifies referring to them as Cubic Equations of State (CEOS).

The first form of CEOS was introduced by Van der Waals in 1873 (Pedersen and Christensen, 2007) . His equation is based on a modification of real gas law. It approximates the behavior of real fluids by taking into account the nonzero size of molecules and the attraction forces between them. The nonzero size and the attractive parameters are represented by parameters ***b*** and ***a*** respectively. These make the equation to be referred to as two-parameter cubic equation of state.

Redlich and Kwong (1949) observed that Van der Waals equation was very weak in predicting vapor pressure and liquid phase molar volume. They pointed out that the weakness in vapor pressure prediction is because Van der Waals ignored the temperature dependence of attractive term (parameter ***a***) in his equation. In this light they added temperature in the attractive term formulation and adjusted other parameters to improve vapor pressure prediction. Their equation is considered by many as the first modern cubic equation of state (Valderrama, 2003).

Kaback (1976) noted that some authors proposed a parameter to be added to the attractive term of original Van der Waals-type equations in a bid to correct for the effect of the molecular size and shape. He reported that this correction has significantly improved prediction of the properties of non-polar fluids. He also acknowledged that the most widely used and most reliable amongst these parameters is the Pitzer acentric factor or simply acentric factor. This is the major innovation in the nowadays widely accepted improved forms of the Van der Waals equation of state. Although they incorporate the acentric factor, these equations are still widely known as two-parameter equations of state.

Soave (1972) later found the pure-component vapor pressure estimated by the Redlich-Kwong equation, to still be somewhat inaccurate and therefore suggested to increase the temperature dependency of parameter ***a***. In this respect he expressed parameter ***a*** as a direct function of temperature. He also included acentric factor in its calculation. The resulting new equation was named Soave-Redlich-Kwong (SRK).

[3] The principle of corresponding states proposes that a given property for any fluid component can be related to that of a well-known reference compound (CH_4) at the reduced sate (Pedersen and Christensen, 2007)

One of the inherited deficiencies of the SRK equation is its inability to estimate liquid density with acceptable accuracy. Peng and Robinson (1976) in their original equation related this shortcoming to the fact that Soave-Redlich-Kwong predicts pure components' critical compressibility factor around 0.33, which is higher than measured values for most components (between 0.25 and 0.29). They proposed a new equation of state that predicts critical compressibility factors of pure components closer to their measured values (~0.30). This equation called PR76 improves pure components liquid density estimation as compared with SRK. However the deviation with measured data remains quite significant. Two years later, Peng and Robinson proposed a modification of their original equation. This resulted in a new equation called PR78. The new equation has a new formulation for the temperature dependency of parameter ***a***, and is used for components with acentric factor greater than 0.49.

Some other authors proposed some forms of modifications of these equations still in a bid to improve their accuracy. Amongst them, Twu et al (2002)'s equation is worth mentioning.

Although many improvements in thermodynamic properties' estimation have been achieved by these CEOS, they still have difficulties to accurately predict liquid densities, especial in the vicinity of the critical points. A way to overcome these difficulties is to introduce a volume correction factor to adjust the volume calculated by the CEOS. This factor helps in reducing the deviation between the observed liquid densities and those calculated by a CEOS. The method widely used in the oil industry is the Peneloux volume shift method (Peneloux et al., 1982) which applies to both PR and SRK CEOS. Nevertheless, more complex volume correction methods have been proposed amongst which Mathias and Neiheri (1989)'s is worth considering. This method is limited only to the PR CEOS.

One of the major shortcomings of the classical volume translation models is that they improve only accuracy of the volume estimation, while other properties such as enthalpy and entropy remain with large errors. This is because the translation parameter is neither temperature nor density dependent. Observing this, Frey et al (2007) suggest creating a temperature and density-dependent volume correction for robust and accurate estimation of thermodynamic properties.

Volume corrected cubic equations of state generally provide a good estimation of pure components and mixtures' density, especially when adequately tuned and fit to measured data. Boyle and Carroll (2001) calculated acid gas (CO_2 and H_2S) densities with different equations of state. They arrived at the conclusion that the Peng-Robinson equation of state was satisfactory in computing CO_2 density. Their findings were confirmed by Pusch et al (2010) and Singhe and Ganzer (2010) in their study on the ability of the CEOS to predict CO_2 density. However the accuracy, albeit sufficient for engineering purposes, is still below that of the Span and Wagner's fundamental equation especially in the vicinity of the critical point.

The fact that the critical compressibility factor takes on fixed values in the two-parameter CEOS constitute a major drawback to their accuracy. Consequently, three-parameter equations, where the critical Z-factor is substance-dependent and adjustable, have been developed in a bid to overcome this drawback (Cismondi and Mollerup, 2005, Valderrama, 2003) . One of the early contributions in this respect is that of Patel and Teja (1982). A more recent work has been done by Cismondi and Mollerup (2005). They combined the RK and PR equations to obtain a double parametization of the attraction term (***a***) that introduced a third parameter in the equation.

Orr (2007) observed that once temperature, pressure and composition of a fluid are set, all other properties can also be determined. Therefore once the density is calculated at specified pressure, temperature and composition, the other thermodynamic properties can be estimated with appropriate thermodynamic functions and the Maxwell relations.

As an alternative method for thermodynamic properties calculation via a CEOS, Hagoort (2005) proposed a numerical procedure to compute enthalpy, heat capacity at constant pressure and Joule-Thomson coefficient, as entries in a thermodynamic properties' table. The rationale behind it is the relationship amongst the enthalpy, the isobaric heat capacity and the Joule-Thomson coefficient earlier identified by Klotz (1964). The enthalpy is calculated as aggregate of two terms: the ideal gas enthalpy and the residual enthalpy. The residual enthalpy is a function of the Z-factor, which is calculated with an appropriate equation of state. The other properties are function of enthalpy. Hagoort went on to introduce two numbers phi and omega. Phi is dimensionless and controls the effect of gravity. Omega has the dimension of the square of temperature divided by the square pressure and controls the effect of frictional losses. These numbers are fluid specific and vary with temperature and pressure as such they are also thermodynamic properties (see Figure 2.2). These numbers provide a better control of the performance of the model and ease analysis of the results.

Although Hagoort presented above-mentioned procedure to calculate thermodynamic properties and the numbers Phi and Omega for natural gas in a production well, the same procedure and analogous numbers are derived for CO_2 in an injection well in this work. This enables to extend the thermodynamic properties tables to properties that are not directly estimated by the equations of state. It also permits better model's performance control and easy analysis of the results.

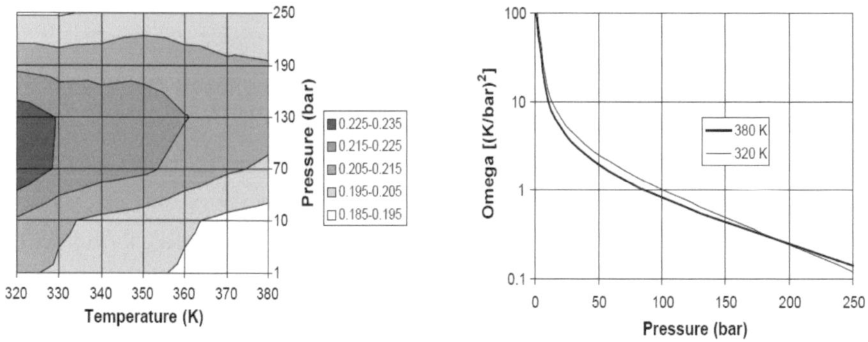

***Figure 2.2**: Psi (left), omega (right) as a function of temperature and pressure in a dry gas (Hagoort, 2005)*

2.3.2 Transport Properties

Transport properties of fluids are known to be directly dependent on temperature in general. In the case of compressible fluids, indirect dependency on pressure through the density has also been recognized.

Fundamentals-based models to calculate transport properties are very complex and generally, the underlying theory is not fully developed (Nito De Castro and Wakeham, 1996). Correlation and hence, prediction of transport properties therefore rely on semi-theoretical and empirical methods. In this respect, custom has been in the petroleum industry to estimate these properties with empirical correlations in the light of theoretical developments.

Nieto de Castro and Wakeham (1996) recognized that the dependence of transport properties on temperature and pressure is not a straightforward one. It can rather be expressed as a sum of three contributions: The dilute-gas property contribution, the excess property contribution and the property critical enhancement.

Mehrotra et al. (1996) made an extensive review of the widely known and accepted models for viscosity prediction and correlation applied to hydrocarbons and petroleum fluids. They classified the models into two broad categories namely semi-theoretical and empirical. This classification was also earlier proposed by Nieto de Castro and Wakeham (1996) for all types of fluids. Some of these models have been used by Zéberg-Mickelsen et al. (2004) in their works. These authors further classified them based on their theoretical background. They distinguished the Self-referencing models, the hard sphere models (Enskog), the Lohrenz-Bray-Clark (LBC) models, the free–volume models and the friction theory models respectively.

Mehrotra and his co-workers went further to highlight that the gas phase viscosity is primarily a function of momentum transfer by translation of the molecules with relatively few collisions. In contrast, the momentum transfer in dense gases and liquids is dominated by collisions and interacting force fields between densely packed molecules. As a consequence, the theoretical description of liquids appears to be difficult due to intermolecular forces as earlier observed by Sheng et al. (1989). These forces are caused by short range (repulsion and hydrogen bonding), wide range (electrostatic) and long range (attraction) effects. Noteworthy is that all the above authors recognized that it is difficult to come up with a widely accepted simple theoretical method for dense gas and liquid viscosity. Sheng et al (1989) further highlighted additional difficulties involved in the study of transport properties. They identified that these were associated with the complexities to perform accurate measurements on one hand and those underpinning the theoretical treatments on the other.

The Lohrenz-Bray-Clark (LBC) model presented by Lohrenz et al.(1964) is the most widely used in the petroleum industry. It is an extension of the correlation of Jossi et al (1962) to mixtures. It is based on residual viscosity concept in line with the general formulation of transport properties given by Nito De Castro and Wakeham (1996). Al-Syabi et al. (2001) proposed a modification of LBC in which they introduced structural and thermal effects. As a result, new correlation coefficients have been generated distinctively for methane alone and for other components. They indicated that mixing rule for the dilute gas viscosity is the one proposed by Herning and Zipperer (1936) . For the other mixture parameters, they used the weighted average method, to reflect the amount of each component in the mixture. They reported to have achieved significant improvement in viscosity calculation.

The estimation of the viscosity with Jossi-based correlations as well as with other correlations requires calculation of the density of the substance. This is more practicably achieved through a cubic equation of state. Therefore, there is the need for estimation of the viscosity and/or other transport properties via equations of state in order to predict thermodynamic and transport properties with a unique model. In that respect, some workers geared their efforts towards estimating viscosity and thermal conductivity with cubic equations of state. In so doing, Sheng et al. (1989) combined the Hard Sphere Theory (Enksog) with the original Peng-Robinson EOS to estimate viscosity and thermal conductivity. Fan and Wang (2006) and Guo et al (1997) recognized the similarities between PVT(Pressure, Volume, Temperature), TμP (temperature, Viscosity and Pressure) and TλP (Temperature, Thermal Conductivity and Pressure) relationships. In the light of these similarities they proposed analogue cubic equations of state for viscosity and thermal conductivity. They reported improved results as compared with those of widely accepted correlations.

Thermal conductivity like viscosity is mostly predicted based on empirical and semi-empirical correlations. Mathias et al (2001) pointed out that quantitative correlations for thermal conductivity of pure fluids and their mixtures require a separation in background contribution and critical enhancement. They proposed a model which used the Mathias and Parekh (1996) model to calculate the background term based on the extended corresponding states principle, which was earlier presented by Ely and Hanley (1983). For the critical enhancement, they suggested the use of a new phenomenological model.

Chung et al (1988) developed density dependent correlations for viscosity and thermal conductivity of non-polar and polar fluid systems. They reported to estimate viscosity and thermal conductivity of investigated fluid systems with average absolute deviation of 4% and 6% respectively.

Some efforts in calculating thermal conductivity with cubic equations of state have also been documented. Worthwhile noting are the works of Sheng et al. (1989) and Guo et al. (2001), earlier mentioned in the foregoing paragraphs on viscosity models.

While recognizing the significance of the forgone efforts in estimating thermal conductivity and viscosity, it is important to highlight that the estimates have always been unsatisfactory. As such the viscosity, would always require matching the models to measured data to obtain a tuned model to be used in simulation. The measured data will in general not cover the whole range of temperature and pressure encountered in the wellbore. As such the tuned model may also estimate the property with large error outside the temperature and pressure ranges used for tuning. In this light it would be judicious to also create tables of viscosity and thermal conductivities using multiparameters equations of state. The values of these properties are read from these tables during simulation and hence ensuring improved accuracy.

2.4 Thermodynamic and Transport Properties of CO_2

Bahadori and Vuthaluru (2009) have recognized that commercial codes and general purpose equations of state are convenient and easy to use in predicting properties of fluids. Ülker (2009) confirmed this view in her dissertation. She investigated CO_2 density and solubility in brines and pure water using ECLISE-PVTI[4] and based on CEOS. Nonetheless Bahadori and Vuthaluru (2009) highlighted the inability of the commercial codes to accurately estimate all the fluids' properties in the entire fluid regions (the gaseous, liquid, near-critical and supercritical regions). Consequently, Bahadori et al. (2009) developed a correlation to predict CO_2 aqueous solubility and density. The method accurately estimates density in the pressure range of 25 to 700 Bar and temperature range of 293 to 433 K.

McCollum and Ogden (2006) also proposed a simple correlations for CO_2 density and viscosity estimation. They reported their accuracies to be acceptable in the range of temperature and pressure relevant for most CCS[5] projects.

Although the above correlations would work well in CCS projects, it would be necessary to have an equation of state that does not only estimate density more accurately in a wider range of temperature and pressure but can also calculate other thermodynamic properties. Span and Wagner (1996) proposed an equation of state in this respect. The equation is explicit in the Helmholtz free energy. It can represent even the most accurate data with experimental uncertainty over the fluid regions from critical points up to 1100 K and 800 MPa. Unfortunately its complexity has always precluded its implementation in commercial codes. Therefore it is often used as a benchmark to these codes, which are usually based on simpler equation of state and correlations.

Quoting Bahadori (2008) and Civan (2008), Bahadori and Vuthaluru (2009) propounded that equations of state are not especially suitable to estimate transport properties. They further highlighted that, Chung et al. (1988)'s method has been extensively used to estimate transport properties of CO_2 in the moderate pressure region, and that the extension to the high pressures has been based on the method of Reid et al. (1987).

Many authors amongst whom Sastri and Rao (1999) observed that theoretical and semi-theoretical correlations on which transport properties estimation are generally based, require density to be calculated from an equation of state. Realizing that the equation of state estimation of density is often in error, it is obvious that the transport properties estimation will inherit additional errors from the density. These inherited errors will therefore result in larger errors in the prediction of transport properties. This is the main weakness of semi-theoretical and theoretical correlations. This remark has prompted Bahadori and Vuthaluru (2009) to come up with a simple density-independent correlation for the

[4] Commercial PVT simulation software proprietary of Schlumberger
[5] Carbon Capture and Sequestration

estimation of viscosity and thermal conductivity of CO_2. They reported results to deviate from measured data by 2% and 2.5% for viscosity and thermal conductivity respectively. Earlier efforts of McCollum and Ogden (2006) were also in this direction, albeit limited only to viscosity.

Notwithstanding, some models that use density as an independent variable have been reported to yield results of high accuracy. These models take advantage of the accuracy of multi-parameters equations of state to calculate density. Consequently, they are very complex and therefore, mostly used as standard for benchmark to computer codes based on simpler earlier- mentioned correlations. Amongst them, Vesovic et al (1990) presented equations for viscosity and thermal conductivity using density estimated with the equation of Ely et al (1987). Their equations are reported to be based in part on experimental data that have been critically assessed for their consistency and for their agreement with theory whenever possible. These equations are also based on theoretical correlations to complement the absence or inaccuracy of experimental data. The complete models cover the temperature range of 200 K to 1500 K for viscosity and 200 to 1000 K for thermal conductivity, and pressures up to 100 MPa. The uncertainties associated with these models vary according to the thermodynamic state from ±0.3% for the viscosity of the dilute gas at near room temperature to ±5% for the thermal conductivity in the liquid phase.

Fenghour et al (1998) revisited Vesovic et al's viscosity equation based on new measurements. They improved its accuracy in the liquid region. They also extended the range of applicable pressures to 300 MPa. However limitations owing to the equation of state that they used in calculating density restricted the applicability of their equation to 30 MPa at high temperatures.

In order to come up with a new correlation for thermal conductivity, Scalabrin et al. (2006) used Span and Wagner (1996) equation to calculate density. They also implemented the same optimization techniques as Span Wagner. They reported accuracy to be slightly better than that of Vesovic et al.'s equations. In fact the accuracy of Span and Wagner's equation in determining density corroborates to the fact that very little error will be inherited from density. The optimization technique yields a model that is clearly very complex to implement in computer programs. Therefore this equation also majorly serves as a benchmark to commercial codes. It is valid for temperatures from triple-point to 1000 K and for pressures from atmospheric to 200 MPa pressure.

In the light of the foregone paragraphs, the high accuracy of the MPEOS is an advantage that should not be neglected in predicting thermodynamic and transport properties of the flowing fluids in a wellbore. This is especially very relevant for CO_2 because it would transit into liquid and near critical regions, where predictions with conventional CEOS are often in high errors. The efficient use of the MPEOS can be possible without seriously affecting their accuracy. This entails using them to create tables of properties from which the simulator reads the data. This avoids high computation time caused by invoking the equations at each wellbore segment. It also simplifies integration of these equations into simulation codes. As such this work adopts this procedure by integrating runtime thermodynamic and transport properties table in the codes. The thermodynamic properties are created with the Span and Wagner equation, the viscosity table with the Fenghour et al. equation and the thermal conductivity with the Scalabrin et al equation.

2.5 Vapour Pressure

Many authors presented correlations to estimate vapor pressure of pure components. The first and simplest correlation is reported to be Antoine's equation, developed in 1888 (Edalat et al., 1993). It was derived from the Clasius-Clapeyron relation and describes the relationship between vapor pressure and temperature. Edalat et al (1993) recognized that one of the weaknesses of this equation is that the same values of its coefficients cannot be used in all fluid regions, from triple-point to critical point. Consequently, the use of this equation requires development of a large data bank of coefficients over the fluid regions of interest. However, this weakness has been overcome by extending the equation with additional parameters.

Later on, Wagner (1973) proposed a generalized equation with four parameters, which are characteristic constants of the fluid. The accuracy of this equation is acceptable only up to reduced temperature of 0.5.

Some other efforts have been consecrated on developing vapor pressure equations with better accuracy and covering the entire region from triple-point to critical point. In that respect, CHERIC (2010) also proposes correlations to estimate temperature dependent properties of pure components including vapor pressure. Unlike Antoine's equation, CHERIC's vapor pressure equation uses the same value of parameters over fluid region of interest. These parameters were developed through curve-fitting of measured data.

Some authors did not only develop equations that cover the entire vapor pressure region but they also introduced acentric factor in a bid to improve accuracy. Amongst them, Lee and Kesler (1975) are worth mentioning. Although the prediction of their equation is pretty accurate, it performs poorly for pressures below one (1) atmosphere, where deviations are in excess of 10%.

Much later, the limitations of Wagner's equations and the importance of acentric factor were also recognized by Edalat et al. (1993). He modified Wagner's equation by establishing its four parameters as a function of the fluid's (including mixtures) acentric factor. Consequently, their equation does not only yield more acceptable results up to critical point, but it can also easily be extended to mixtures.

The general trend in the petroleum industry has been to estimate vapor pressure with cubic equations of state to establish vapor-liquid equilibrium based on equality of phase fugacities. This has been pushed firstly by the need to use a unique equation for all thermodynamic properties' calculation and secondly by the necessity to extend the predictions to mixtures. Many authors proposed cubic equations of state to closely reproduce vapor pressure of pure components (Peng, 1976, Soave, 1972, Twu, 2002) . These equations are quite successfully applied to multi-component fluids Vapor-Liquid Equilibrium (VLE) calculation when using appropriate mixing rules (Twu, 2002).

Specifically related to CO_2, Span and Wagner (1996) developed an equation that calculates carbon dioxide's vapor pressure with uncertainty estimated to be of maximum 0.012%. The correlation was developed on the basis of measured data from Duschek et al. (1990b). The drawback of this equation is the difficulty to directly implement it into a simulation code on the one hand and the fact that it will consume a lot of CPU-time on the other.

Although most of the above-discussed correlations are of acceptable accuracy, the superiority and the wide coverage (in terms of temperature and pressure ranges) of Span and Wagner's equation cannot be over-emphasized. Therefore it is used in this work to calculate vapor pressure. As with the other thermodynamic properties a vapor pressure table is created for the purpose. The simulator reads the value at prevailing temperature and compares it with prevailing pressure to identify if the phase of the fluid is changing in the well.

2.6 Phase Equilibrium

Effective and accurate phase equilibrium calculation is very important in modeling and simulating multi-phase flow processes (Voutsas et al., 2004). Two fundamental modeling approaches have been extensively discussed in the literature: the activity coefficient modeling (based on chemical potential) and the fugacity coefficient modeling approaches (Louli et al., 2007, Orr Jr, 2007). Walas (1985) provides guidelines to select the right model for process simulation. One of his important recommendations is that, when the reduced temperature is less than 0.75 the two approaches should be adopted. In this way the activity coefficient approach is used for the liquid phase and the fugacity coefficient counterpart for the vapor phase.

Atkins (1986, P.167) recognized that ideal mixtures obey Raoult's law (equation 2.6.1) .

$$\chi_i P_T = x_i P_{Vi} \qquad 2.6.1$$

Where,

P_T is the total pressure of the system

P_{Vi} is the vapour pressure of component *i*

X_i and χ_i are mole fraction of component *i* the liquid and vapour phases respectively.

These mixtures are those containing components which are chemically similar, therefore there is very little or no interaction amongst their molecules. As a consequence, Lolley and Richardson (1998) highlighted that the K-values of individual components in a mixture, depend strongly on temperature and pressure and not on composition. They however pointed out that this holds in all fluid regions except in the vicinity of the critical point. They further related the K-values to total pressure and vapor pressure of individual component by equation 2.6.2 below.

$$K_i = \frac{\chi_i}{x_i} = \frac{P_{Vi}}{P_T} \qquad 2.6.2$$

As earlier emphasized, it is custom in the petroleum industry to use advanced cubic equations of state (CEOS) such as Peng-Robinson or Soave-Redlich-kwong to perform VLE based on the fugacity coefficients of the fluid. The CEOS can be applied to mixtures at various temperatures and pressures to yield a consistent description of the fluid. The main challenge is to select appropriate mixing rules for calculations involving mixtures. To this end, Van der Waals mixing rules have shown to work quite well with non-polar[6] mixtures (Soave, 1972, Twu, 2002). The necessity to improve modeling of polar[7] or associating mixtures[8] and mixtures containing molecules of very different sizes, required a more complex description of the molecules' interactions. The simplest approach has been to modify Van der

[6] Non-polar fluids are fluids in which constituent molecules have no polarization

[7] Polar fluids are fluids in which constituent molecules have a polarization e.g. water, alcohols

[8] Associating fluids are fluids showing a chemical effect due to hydrogen bonding and charge transfers which link molecules as monomers to form di-mers. An example is water

Waals' mixing rule by introducing a binary interaction coefficient in the expression of the mixture's attractive term, ***a*** (Orr Jr, 2007, Twu, 2002). More complex mixing rules with varying degrees of success have also been reported (Twu et al., 1998, Voutsas et al., 2004, Valderrama, 2003) . Of great importance is the universal mixing rule proposed by Voutsas et al (2004). It applies to all types of system asymmetries. This mixing rule uses the UNIFAC[9] binary interaction parameters, derived from excess Gibb's free energy and introduced by Fredenslund et al. (1975). Louli et al.(2007) have combined Voutsas et al.'s mixing rule with Peng-Robison CEOS to predict phase behavior of synthetic natural gas and oil systems. They noted that this type of mixing rules have been gaining ground in the past two decades. The resulting models are referred to as equation of state with Gibb's free energy mixing rule (EOS/G^E). McDonald and Floudas (1995) recognized another important advantage of using UNIFAC group contribution methods. Unlike simple binary interaction parameters methods, these models offer the possibility of predicting the presence of more than one liquid phase.

Fundamental approaches to vapour-liquid equilibrium require optimization algorithms that minimize Gibb's free energy in the system (McDonald and Floudas, 1995, Pan and Firoozabadi, 1998). This requires complex algorithms for both stability and phase equilibrium calculation. McDonald and Floudas (1995) argue that the conventional optimization methods based on Newton techniques do not always guarantee convergence to global minimum. Consequently, they employed an algorithm that uses a convex formulation of the minimization problem. Later on, Pan and Firoozabi (1998) implemented simulated annealing to perform multiphase calculations on petroleum reservoir fluids. The draw-back of these methods is their complexity and CPU-time required for the calculations. These make them to be less attractive for practical applications.

The convention in the petroleum industry as recognized by Orr Jr. (2007), is to apply fugacity coefficient models in flash and negative flash calculations associated with stability tests.

This work is limited to pure CO_2 of which an example phase diagram is shown in Figure 2.3 below. As such phase equilibrium does not require selection of I binary interaction coefficients for CEOS. In general this requirement would force preference (to the detriment of MPEOS) on the fundamental approach based on minimization of Gibb's free energy. Furthermore the complexity of the optimization algorithms used with the multi-parametric equations of state as well as their simulation time-intensiveness can be avoided by creating thermodynamic properties tables from which a simulator will

[9] Universal Quasi-chemical Functional Group Activity Coefficients

read the K-values. This approach has been chosen for this work. The vapour-liquid equilibrium is based on the K-values generated with the fundamental equation of state of Span and Wagner.

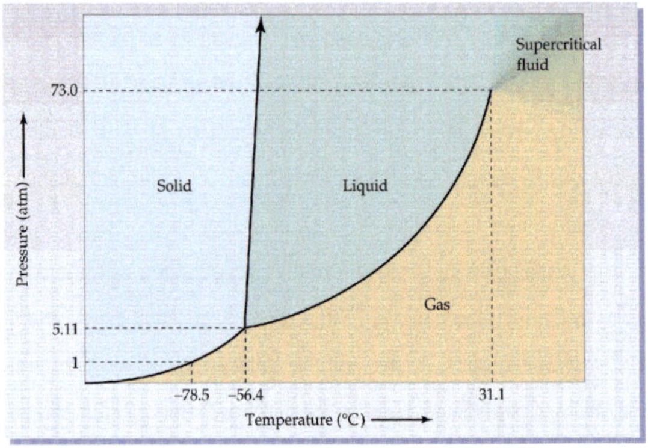

Figure 2.3: *Example of CO_2 Phase diagram (http://www.teamonslaught.fsnet.co.uk)*

2.7 Heat Transfer

The transfer of heat between the flowing fluid and the surrounding formation and vice-versa, will have an important influence on the temperature and hence the thermodynamic and transport properties of the flowing fluid in the wellbore. The mechanisms by which heat flows in the transfer process can be classified as convection (both forced and natural), radiation and conduction. Forced convection is the dominant phenomenon in the flow conduit, where heat will be transferred from the flowing fluid to the conduit's wall and vice-versa. Natural convection is the major mode in an annulus filled with a stationary fluid. Radiation may also be pronounced in the annulus depending on the prevailing temperatures in the inner and outer wall of the annulus. The transfer of heat between casing, cement, formation and within the formation is dominated by conduction. Figure 2.4 gives a schematic representation of the temperature distribution as result of the heat transfer process in the wellbore segments and the surrounding rock.

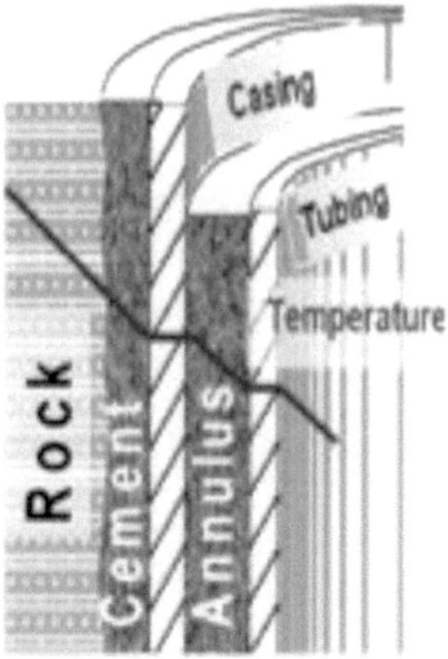

Figure 2.4: *Temperature distribution as a result of heat transfer process in a segment of the wellbore and the surrounding rock (DBI, 2011)*

Earliest work on modeling transfer of heat between the fluid in the wellbore and the surrounding formation include that of Ramey(1962) who proposed an analytical solution to estimate temperature of hot water in the tubing during injection, under the assumption that the wellbore is an infinite line source. The solution considers steady-state heat transfer in the wellbore and unsteady-state transfer in the surrounding formation. The latter is represented by a transient heat conduction function $f(t)$. The approach enables estimation of temperature of fluids and that of tubing and casing as well as that of the surrounding rock as a function of depth and time.

Ramey also proposed a procedure to calculate heat transfer coefficient and stressed on the importance of its accurate estimation on the accuracy of his temperature model. In this light, Willhite (1967) further developed Ramey's idea and proposed a model to calculate an overall heat transfer coefficients for the wellbore (from formation-cement interface to fluid in the tubing) in steam and hot water injection wells. His model yielded acceptable results. It assumes transfer to be mainly by conduction in tubing, casing and cement rings and by both radiation and convection in the annulus. The calculation follows an iterative algorithm earlier proposed by Satter (1965), who investigated the heat losses due to cooling and condensation during flow of steam down a wellbore. His work was an extension of the Ramey's model to account for phase change and associated heat loss/gain in a steam injection well.

Literature on heat transfer in general and heat transfer coefficient in particular is rather exhaustive in mechanical and chemical engineering fields. Coulson and Richardson (1996, P.382) recognized that the calculation of heat transfer coefficient is a key requirement in any design problem involving heating or cooling. They described heat transfer coefficient as the indication of the ability of a medium to transmit heat by referring to its reciprocal as heat resistance. They further presented the different processes of heat transfer as a function of the medium and recognized that the temperature gradient is the driving force for heat transfer. They further recognized that heat transfer by conduction is the main mode taking place in solid bodies, while convection, both natural and forced are predominant in fluids. They went on to emphasize the importance of radiation in the process of heat transfer through gases. They also pointed out that radiation is the only means of heat transfer in a vacuum.

Ramey's approach was based on the line source concept implying that the well is a cylinder of infinite length in infinite medium (Alves et al., 1992, Durrant and Thambynayagam, 1986, Hagoort, 2004). It was also based on perfect gas behavior and considered no phase change in the flow conduit. Eppelbaum and Kutasov (2006) highlighted the similarities between Darcy's and Fourier's laws. In the light of these similarities they identified that the same diffusivity equation describes the transient flow of incompressible fluid in porous media and the transient heat conduction in solids. This was also pointed out by Brill and Mukherjee (1999, P.15) who highlighted that the $f(t)$ function in Ramey's model can also be expressed with the exponential integral function like the pressure diffusivity equation. Consequently, $f(t)$ function's solution can also be approximated by a logarithmic function for times greater than one week. This raises the problem of validity of the Ramey-type solutions for early time period smaller than one week. Hagoort (2004) associated this problem to the fact that Ramey's solution assumed negligible accumulation of heat and ignores the effect of temperature history. Therefore, the principle of superposition would not hold. He argued that this holds only for temperature functions in the form of a step function, or for time-independent heat-loss functions. He further noted that these assumptions would be approached only after some time when both heat losses and temperature become slowly-varying functions of time.

Notwithstanding its shortcoming, Ramey's model has been widely adopted. Many authors applied it with additional features. Amongst them, Alves et al (1992) proposed a unified model for multiphase flow in pipes of all inclinations. This model relies on a rigorous treatment of thermodynamic behavior of the flowing fluid similar to the approach that Coulter and Bardon (1979) used in modeling temperature distribution in pipelines. Consequently, it degenerates into Ramey's equation for the flow of ideal gases and incompressible liquids. It also reduces to Coulter and Bardon's equation, when steady state heat transfer and constant temperature are assumed in a horizontal pipe.

Improvements of the accuracy of Ramey's solution at early times have been proposed by Hasan and Kabir (1991). They included the superposition principle in the Fourier's law of heat conduction in order to account for the gradual change in heat transfer rate between wellbore and formation. They proposed two expressions for $f(t)$: one which is valid for early time (dimensionless time less than 1.5) and the other one valid for later times. The discontinuity of the above $f(t)$ function at dimensionless time equal to 1.5, urged Hasan and Kabir (1994) to later suggest a continuous form $f(t)$.

Hagoort (2005) also proposed an analytical model to predict temperature in a gas production well. The model is simple and based on dimensional analysis. This yields dimensionless numbers, which are used to demonstrate the improved physical transparency of the approach. The model is based on extended Bernoulli's equation and involves a simultaneous estimation of temperature and pressure. The pressure solution is the analytical solution developed by Lingen (1974).

Eppelbaum and Kutasov (2006), considered the wellbore as a cylindrical heat source and presented a semi-analytical solution to compute the wellbore temperature. This solution can be rearranged to get an expression for $f(t)$ for which, the principle of superposition applies without limitations. Therefore the equation holds for all time periods. This approach is of great importance in temperature well testing, where early time heat conduction plays a significant role.

Apart from the Ramey-type analytical approaches, some other approaches with diverse success and complexities were also presented. Wu and Pruess (1990) proposed an analytical solution for wellbore heat transmission in layered formations with solution found both by Laplace transformation and in the real domain. They also provided a more accurate formula for the transient heat conduction function $f(t)$. The solution gives very accurate results in all time periods. Numerical solutions have also been reported in the literature. Wooley (1980) and Farouq Ali (1981) 's works are worth mentioning. They are more comprehensive and consider both horizontal and vertical heat conduction in the formation and are applicable to a wide range of operations' conditions. Unfortunately, advantages of these models have not been usually utilized. They are too complicated for field applications and simulation studies, because they require a lot of parameters which are seldom known with precision (Wu and Pruess, 1990).

Bahonar et al. (2010) investigated the implications of steady state heat flow assumption in a steam injection well. They recognized that assumption of steady-state heat transfer in tubular and cement is reasonable for long time processes. They compared three $f(t)$ functions and confirmed the fact that the Ramey's model worked very well from time period greater than one week. They also found out that the Fontanilla and Aziz (1982) and Hasan and Kabir (1994) models hold for all time periods. Fontanilla and Aziz's model increases in accuracy as time gets larger; whereas that of Hasan and Kabir's model becomes less accurate as time increases. They finally recognized that assumption of steady-state heat transfer in tubular and cement becomes more reasonable as injection time increases.

2.8 Heat Losses by Condensation and Cooling

Satter (1965) investigated the effect of condensation and cooling and their associated heat losses during steam injection operations. He proposed calculation procedures to estimate heat losses for superheated, saturated and under-saturated steam. He also analyzed the effect of injection rate, time, pressure, temperature and depth on heat losses. His approach to heat transfer with the surrounding is basically the same as that of Ramey (1962). His solution procedure divides the wellbore into segments, where steam has constant thermophysical properties. However it considers different overall heat transfer coefficients that change with temperature and depth.

He accounted for different processes for heat losses, depending on the state of the steam.

1. For the superheated steam: (i) decrease in temperature until steam is cooled to the saturated sate; (ii) Condensation until complete condensation into hot water.
2. For saturated or under-saturated steam: (i) condensation until complete condensation into hot water.

The following assumptions are considered in his model, which addresses flow in the tubing with annulus isolated by downhole packers.

1. Steam is injected at a constant rate, pressure , temperature and quality
2. The heat transfer in the wellbore is steady state while the heat transfer in the earth is unsteady state.
3. Kinetic energy changes are negligible
4. There is negligible variation of thermal conductivity and diffusivity of earth with depth.
5. The thermal and physical properties of steam are constant within given depth interval.

Heat loss due to condensation and/or vaporization is calculated as a percentage of heat loss in the well segment, and the equations differ as a function of the state of the steam.

- For superheated steam at the surface

In case of cooling:

$$\%Heat\ Loss = \frac{\left[H_i - H(Z,t) + gZ\right]}{H_i - (H_w)_{T_m}} \times 100 \qquad 2.8.1$$

In case of Condensation:

$$\%Heat\ Loss = \frac{\left[H_i - \{H_w + y(Z,t)L_v\} + gZ\right]}{H_i - (H_w)_{T_m}} \times 100 \qquad 2.8.2$$

- For saturated or undersaturated steam

$$\%Heat\ Loss = \frac{\left[\{y_i - y(Z,t)L_v + gZ\}\right]}{H_w + y_iL_v - (H_w)_{T_m}} \times 100 \qquad 2.8.3$$

Where,

H_i is the specific enthalpy of steam at initial pressure and temperature

L_v is the latent heat of steam at injection pressure

H_w is the specific enthalpy of saturated water at injection pressure

T_m is the mean surface temperature of the earth

Z is the depth below the surface

g is the acceleration due to gravity

Satter's approach being laudable is however of greater significance for steam injection. This is because latent heat is very high in steam (2260 KJ/KG) compared to CO_2 (855 KJ/Kg). Therefore there would be larger quantities of heat transferred during phase change and as such errors caused by calculating heat of condensation and evaporation as a percentage of heat loss in the well segment will not significantly affect the results. This would not be the case for CO_2. A more fundamental approach based on physics would yield a more accurate estimation. In this respect the quantity of heat transferred due to phase change is estimated in this work based on the enthalpy of evaporation which is computed as a function of entropy change during the phase transition process. The entropies of the two phases are read from the thermodynamic properties tables generated with the Span and Wagner's equation.

2.9 Efforts to Predict Temperature and Pressure of CO_2 in a Wellbore

In general CO_2 injection projects have been planned for the purpose of enhancing oil recovery. Supercritical CO_2 is generally appropriate for this application. This fact simplified modeling efforts as acknowledged by Hein (1985) who identified that CO_2 flows most effectively in its supercritical form. Therefore CO_2 was often injected above its critical point and thus experienced no phase transition in the wellbore. As such conventional models for single phase flow were conveniently used to predict wellbore flow performance. Nevertheless as earlier mentioned Cronshaw and Bolling (1982) earlier presented a numerical model for non-isothermal flow of carbon dioxide in wellbores. They emphasized on the need for accurate estimation of thermophysical properties of CO_2. Their model was validated with pressures and temperature measured in well producing CO_2 with 3% impurities. They reported good model accuracy and went on to examine the sensitivity of their model to wellhead pressure and temperature and flow rate in an injection well.

As also earlier mentioned, more recently Lu and Connell (2008) proposed a numerical procedure to evaluate the flow of carbon dioxide and its mixtures in non-isothermal wells. They solve the heat, mass and momentum conservation equations to provide solutions for temperature and pressure profile. They use Peng-Robinson CEOS to predict thermodynamic properties and phase equilibrium. Unfortunately CO_2 injection data were not available to test their models. They rather used literature data of a steam injection well and a data of production well presented by Cronshaw and Bolling (1982) .

Paterson et al.(2008) later used Lu and Connell's model to simulate CO_2 flow in wellbores. They found out that geological storage of carbon dioxide will often happen at conditions above critical temperature and pressure. However they identified that boiling and condensation could occur in the wellbore. These would particularly happen in the sections close to the surface, where temperatures are below the fluid's critical temperature (31.04°C). To investigate CO_2 behavior in the wellbores of different depths (shallow, deep), they calculated density profiles for wells experiencing different thermal conditions to relate bottomhole pressures to wellhead pressures. They reported that in static columns, for deeper wells, a stable liquid column can exist from the bottom to the top of the well. Contrarily, for shallower wells or wells in depleted reservoirs, static columns can be initially unstable, with two phase conditions near the surface. These findings have been corroborated by Henninges et al. (2010) in an observation well of approximately seven hundred meters depth. They presented a snapshot from a camera inspection video showing boiling liquid phase CO_2 at around 290 meters depth in the well (Figure 2.5). They also presented the condition of CO_2 within the well in a phase diagram (Figure 2.6). Paterson et al further identified that the bottom-hole temperature in an injection well is confined between wellhead and reservoir temperatures.

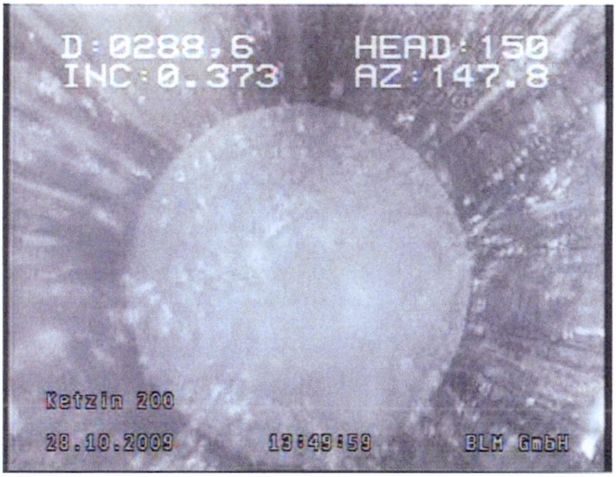

***Figure 2.5**: Snapshot from a camera inspection video in an observation well in a shallow aquifer showing boiling liquid phase CO_2 surface at approximately 290 meters depth (Source: Henninges et al., 2010)*

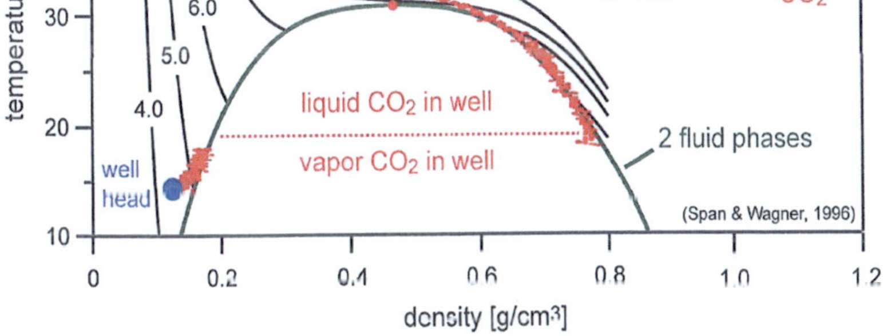

***Figure 2.6**: CO_2 condition in an observation well in a shallow aquifer (Source: Henninges et al., 2010)*

The above points have highlighted that to properly model CO_2 flow in a wellbore it is required to consider a procedure that accurately predicts phase change as well as its effects on temperature and pressure. Such procedure is applied in this work. The phase equilibrium is predicted with very high accuracy by the Span and Wagner's equation. The accuracy of this equations is verified and confirmed in the following chapter where predictions of thermodynamic and transport properties are compared to measured data.

3. PREDICTION OF THERMODYNAMIC AND TRANSPORT PROPERTIES

The accuracy with which thermodynamic and transport properties are estimated is very important in modeling CO_2 flow in pipes in general and in wellbores in particular. In fact, the fluid pressure and temperature along the flow line are strongly dependent on these properties. As earlier introduced tables of thermodynamic and transport properties data are generated with multi-parameters equations of state (MPEOS). Thermodynamic relationships are used to compute the other properties that are not directly calculated with these equations.

The rationale for using MPEOS is that the predictions made by these equations are more accurate than those of conventional CEOS. The improved accuracy is better appreciated when bearing in mind that other thermodynamic properties as well as the transport properties (viscosity and thermal conductivity) depend on density. This dependence implies that minimizing errors on density will also reduce the uncertainties on transport properties.

It is also important to mention that one of the major requirements for tables in which data are interpolated is continuity in their entire range of temperature and pressure. The thermodynamic and transport properties in this work do fulfill this requirement. The plot of the data in the density and the viscosity tables are presented in figure 3.1 and 3.2 below to show the continuity of data in the entire range of temperature and pressure. Similar figures for other properties can be found in the next section and in Appendix B.

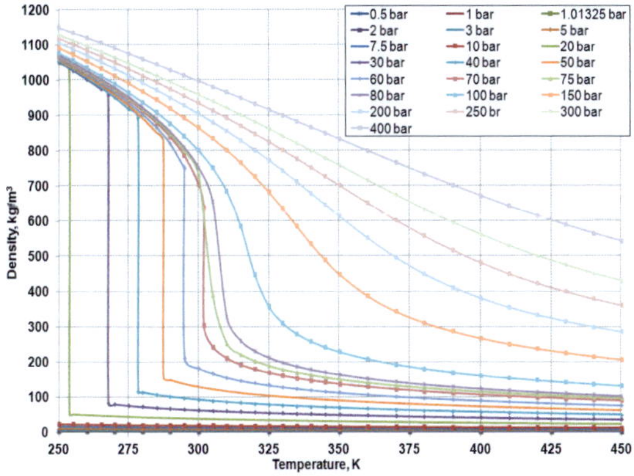

***Figure 3.1**: Isobars of density showing continuity of the entries in the thermodynamic properties table*

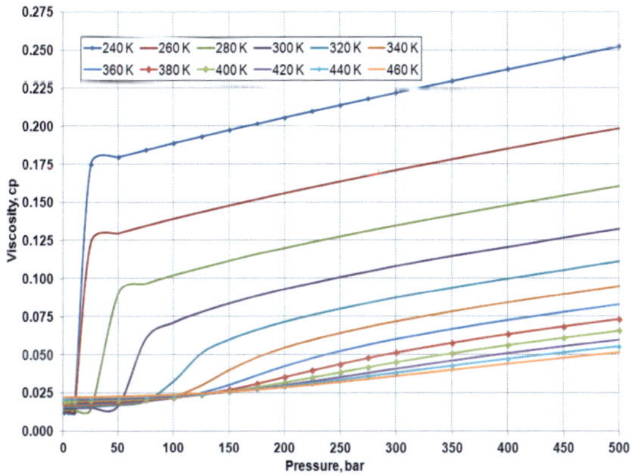

***Figure 3.2**: Isotherms of viscosity showing continuity of the entries in the transport properties table*

3.1 Thermodynamic Properties

As already mentioned in the foregone paragraphs, Hagoort (2005) suggested building thermodynamic properties table in which a wellbore simulator can read values at tabulated temperature and pressure and interpolate to calculate values that fall in between those tabulated. This approach has been used in this work. To ensure high accuracy, the tables are built with data calculated with the MPEOS developed by Span and Wagner (1996). The table entries are extended to include Joule-Thomson (JT) coefficient, volumetric thermal expansion (TE) coefficients, Omega (ω) and Psi (ψ) and K values. The extension is based on thermodynamic functions. Equation 3.1.3 and 3.1.4 show equation to calculate the Joule-Thomson coefficient and Figure 3.3 presents isobars of JT coefficients in the table.

$$\eta_{PiTj} = \frac{1}{C_{P_{PiTj}}} \frac{RT_j^2}{P_i} Z'_{i,j} \qquad 3.1.1$$

Where,

$$z'_{ij} = \frac{Z_{T_{j+1},P_i} - Z_{T_{j-1},P_i}}{T_{j+1} - T_{j-1}} \qquad 3.1.2$$

z is the compressibility factor

R is the universal gas constant

C_p is the isobaric heat capacity

η is the Joule-Thomson coefficient

P is the pressure

T is the temperature

i is index of pressure entry

j is index of temperature entry

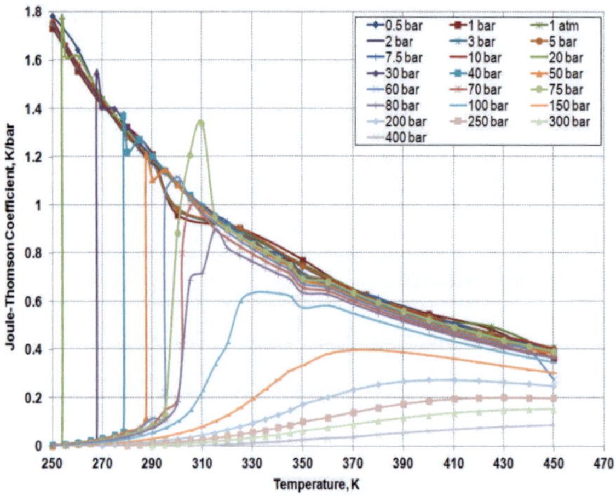

***Figure 3.3**: Isobars of Joule-Thomson coefficient in the thermodynamic properties table*

Similarly Omega and Psi are calculated with equations 3.1.3 and equation 3.1.4 respectively.

$$\omega = \frac{\eta T}{P} \tag{3.1.3}$$

$$\varphi = \eta \frac{P}{T} + \frac{zR}{C_{p_m}} \tag{3.1.4}$$

Where,

z is the compressibility factor

R is the universal gas constant

C_{p_m} is the isobaric molar heat capacity

η is the Joule-Thomson coefficient

P Is the pressure

T is the temperature

Omega is a function of JT coefficient and as such it will equate to zero for an ideal gas. It has the dimension of the square of temperature over the square of pressure (K^2/bar^2 in SI units). It controls the temperature change due to frictional effect. Psi is dimensionless and made up of the addition two terms. The first term is a function of JT coefficient and the second term is a function of the Z-factor and the ratio of the universal gas constant to the molar isobaric heat capacity. Therefore, for an ideal gas, it will equate to the ratio of the universal gas constant to the molar isobaric heat capacity. It controls the change in temperature due to gravity effects. The influence of Psi and Omega on temperature is clearly elucidated in the chapter covering heating in an injection well and analyzed in the chapter on model validation. Figure 3.4 and Figure 3.5 show isobars of Psi and Omega respectively in the thermodynamic properties table.

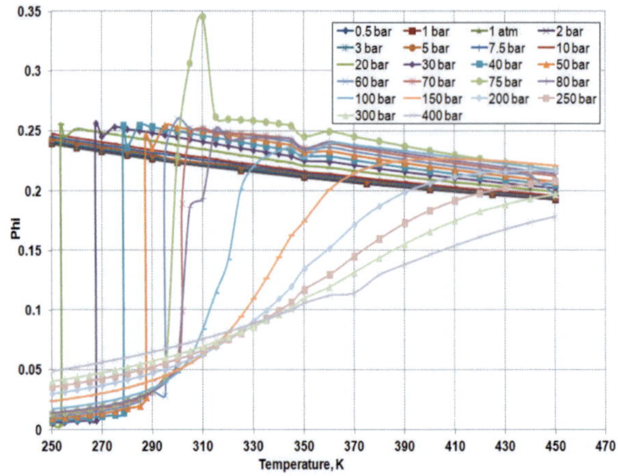

***Figure 3.4**: Isobars of Psi in the transport properties table*

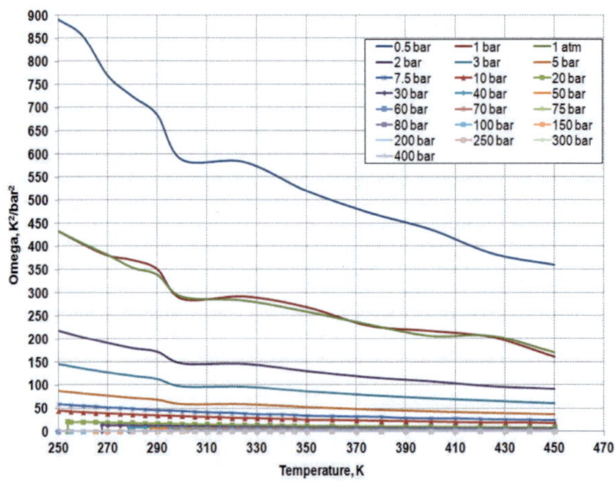

***Figure 3.5**: Isobars of Omega in the transport properties table*

The volumetric thermal expansion coefficient is calculated with equation 3.1.5 and the Joule-Thomson coefficient, which is based on a formula proposed by Hagoort (2005) is calculated with equation 3.1.1.

$$\alpha_v = \frac{V_{i+1} - V_{i-1}}{V_i\left(T_{i+1} - T_{i-1}\right)} \qquad 3.1.5$$

Where,

α_v is volumetric thermal expansion coefficient

V is gas molar volume

i is the row index in the table

The vapour and liquid fractions on the saturation line are calculated with equations 3.1.7 and 3.1.8 on the premises of equality of fugacity. In this respect, the K-values are expressed as the ratio of the fugacity coefficient of the liquid phase to that of the vapor phase (equation 3.1.6). Appendix B.3 presents a step by step development of the equations to estimate fugacity and fugacity coefficients based on the relationship between fugacity and chemical potential summarized by Atkins (1986, Pp. 127-128).

$$K = \frac{\phi_L}{\phi_v} = \frac{\chi}{x} \quad 3.1.6$$

$$y = \frac{K}{K+1} \quad 3.1.7$$

$$x = 1 - \chi \quad 3.1.8$$

Where,

K is the K-value

χ is the fraction of vapor phase

x is the fraction of the liquid phase

ϕ_L is the fugacity coefficient of the liquid phase

ϕ_v is the fugacity coefficient of the vapor phase

3.1.1 Validation of the thermodynamic Properties Tables

The tables are validated by comparing their entries and interpolated values with measured density and vapor pressure values. The measured density data are from Duschek et al. (1990a) and the measured vapour pressure data are from the Dortmund Data Bank Software and Separation Technology's database (DDBST, 2012). The results of the comparison are shown in shown in Figure 3.6 to Figure 3.8. The deviation from measured data is represented with the Absolute Average Relative Error (AARE) and the Maximum Relative Error, expressed by equations 3.1.9 and 3.1.10 respectively.

$$AARE = \frac{\sum_{i=1}^{n} \left| \frac{Measured - Calculated}{Measured} \right| * 100}{n} \quad 3.1.9$$

$$MaxRE = Max\left(AARE_1, AARE_2, ...AARE_n\right) \quad 3.1.10$$

Where *n* is the number of data values.

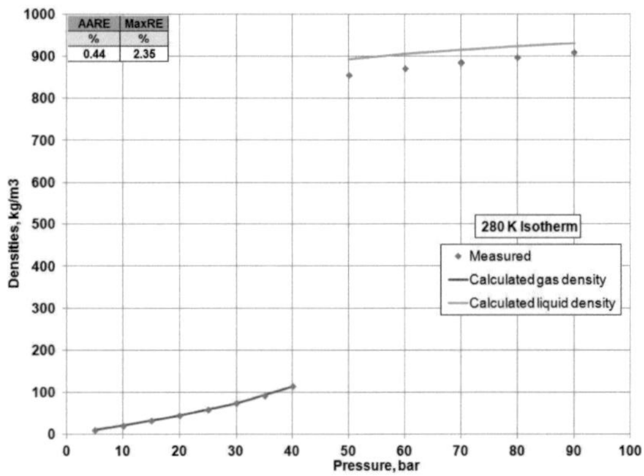

***Figure 3.6**: Comparison of densities' calculated values to measured values at 280K*

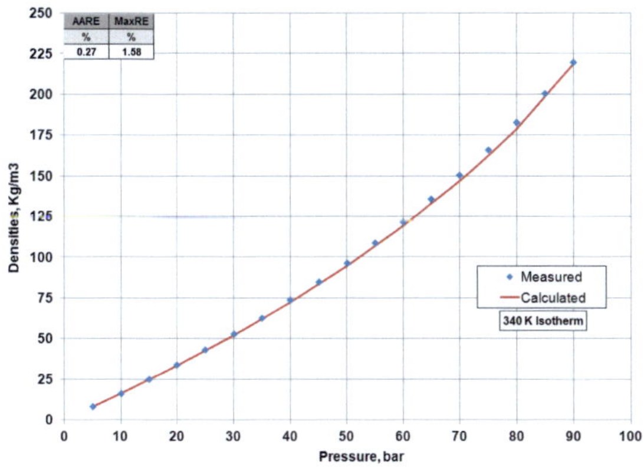

Figure 3.7: *Comparison of density calculated values to measured values at 340 K*

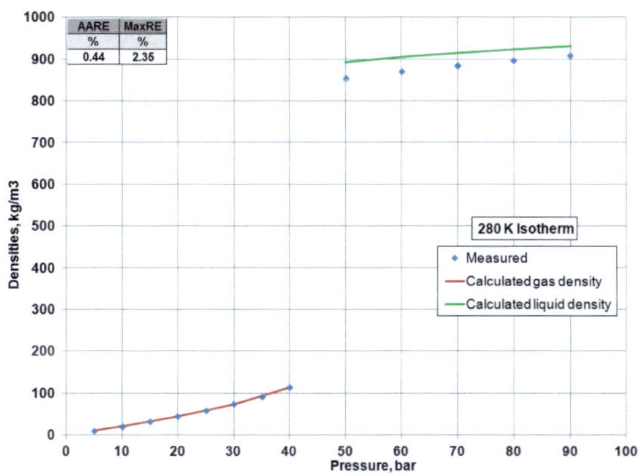

Figure 3.6 and Figure 3.7 show that the deviations between the measured density values and those obtained by reading the table entries and interpolating in-between the entries are not significant. The biggest deviation is around 2.35% and it occurs in the liquid region, where conventional Cubic Equation of State would produce results with significantly higher deviations from measured values. Based on the fact that all the other thermodynamic properties depend on density for their calculation, it can be concluded that the small deviation of estimated density from measured density will enable improve accuracy in the prediction of other properties and therefore that of the pressure and temperature in the wellbore.

Higher accuracy is also obtained for the vapour pressure as presented in Figure 3.8. The maximum deviation is around 0.12%. As a consequence, the improved pressure estimation promoted by high accuracy in density calculation will contribute to improve the accuracy of phase identification in this work.

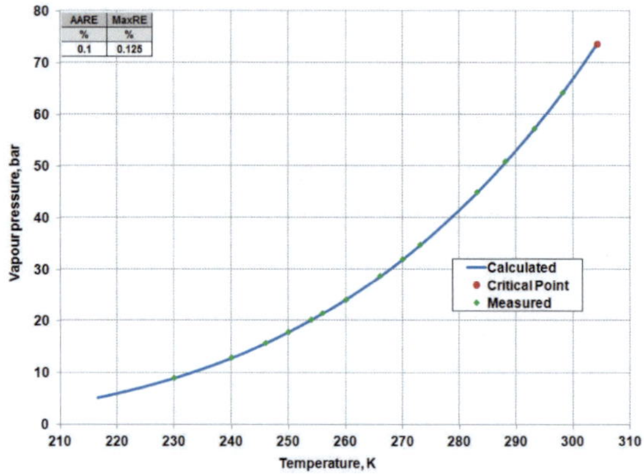

Figure 3.8: *Comparison of calculated vapour pressure values with measured values*

3.2 Transport Properties Tables

The viscosity and thermal conductivity tables are built with reference equations proposed by Fenghour et al (1998) and Scalabrin et al. (2006) respectively. The data cover the range of temperature between 250 and 450 K and pressure 1 and 400 Bar. Special focus is laid on saturation line and near-critical region, where traditional equations of state generally experience difficulties.

3.2.1 Validation of the Transport Properties Tables

The accuracy of the values calculated from the tables is tested with measured data from Pátek et al. (2005) for thermal conductivity and Padua et al (1994) for viscosity. These data are reported to be of high accuracy. Figure 3.9 to Figure 3.11 show how the calculated values of viscosity compare with measured data.

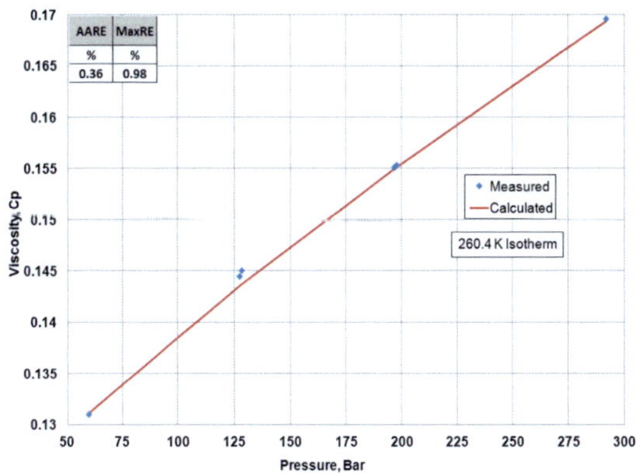

Figure 3.9: *Comparison of calculated viscosity values with measured values at 260.4 K*

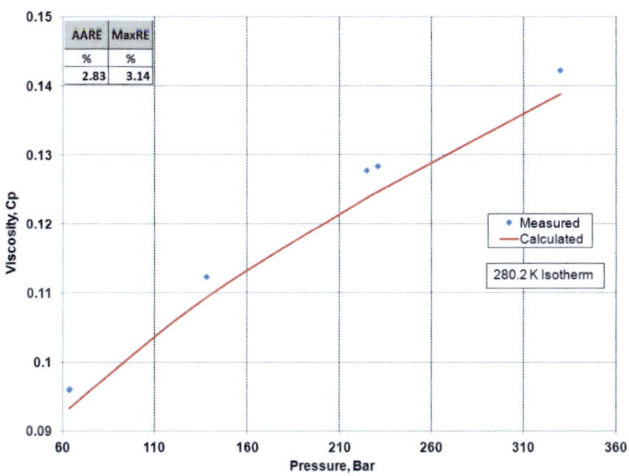

Figure 3.10: *Comparison of calculated viscosity values with measured values at 280.2 K*

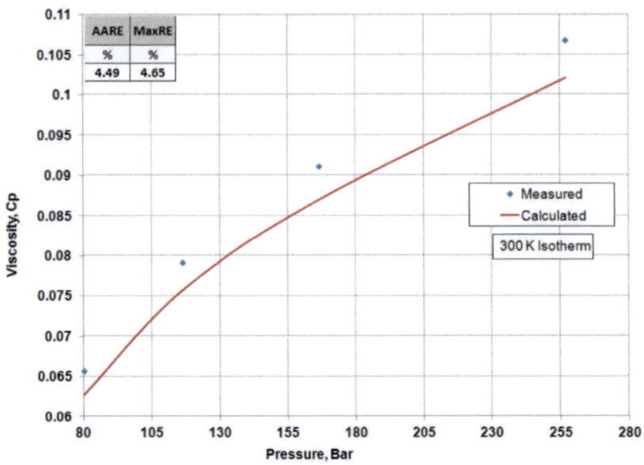

Figure 3.11: *Comparison of calculated viscosity values with measured values at 300 K*

Figure 3.12 presents the comparison of thermal conductivity data read from the table to measured values. The diagonal line represents the line of perfect match where measured and calculated values are exactly the same. The match line represents the actual match and the deviations between the data are reflected by extent to which the match line is removed from the diagonal.

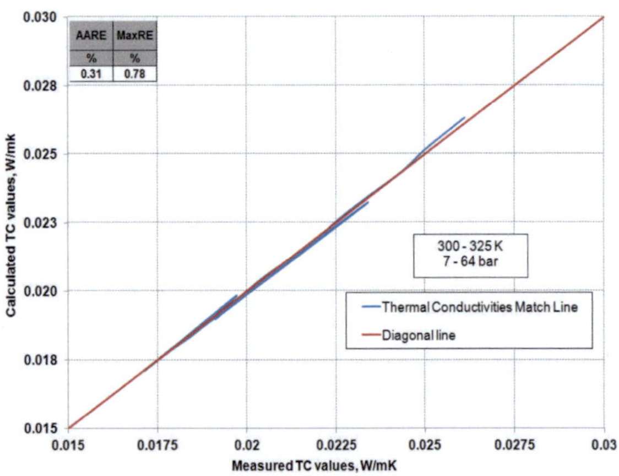

Figure 3.12: *Comparison of calculated thermal conductivities values with measured between 300 and 325 K*

In the light of the above it can be concluded that the estimates from the transport properties tables are very close to the measured values of both viscosity and thermal conductivity. The maximum deviation is less than 5% for viscosity and less than 1% for thermal conductivity. This good match improves the confidence in the estimation of other fluid properties and operational parameters that are dependent on viscosity and thermal conductivity. In this respect, the calculation of the convection heat transfer coefficient between the inside tubing surface and flowing CO_2 as well as that of the frictional pressure loss will inherit very small errors emanating from thermal conductivity and viscosity.

4. INJECTION WELLBORE FLOW EQUATIONS

The flow equations are derived from the fundamental equations for mass, energy and momentum conservation. They result in a system at equilibrium under forces acting on the fluid in a section of the flow conduit as illustrated by Figure 4.1. Since the ratio of length to diameter for the section is large, one dimension flow in the direction parallel to the conduit's length can be assumed.

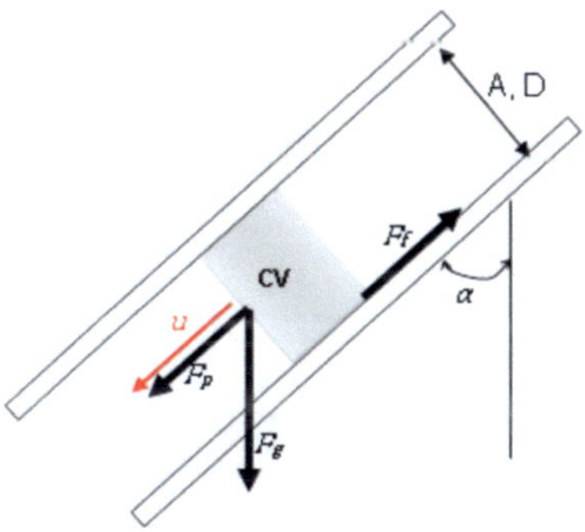

Figure 4.1: *Forces acting on a control volume representing a section of the flow conduit*

4.1 Mass Conservation Equation

Considering the control volume (CV) in figure 4.1, the forces acting on it during injection are the gravity force (F_g), the Frictional forces (F_f), the pressure forces (F_p) and the kinetic forces, which are a result of changes in fluid velocity (u). Application of the mass conservation principle, which Wark and Richard (1999) extensively elaborated upon, implies that there is no accumulation in the control volume at steady state flow conditions. This is illustrated by figure 4.2 and mathematically represented by equation 4.1.1 through 4.1.3 below.

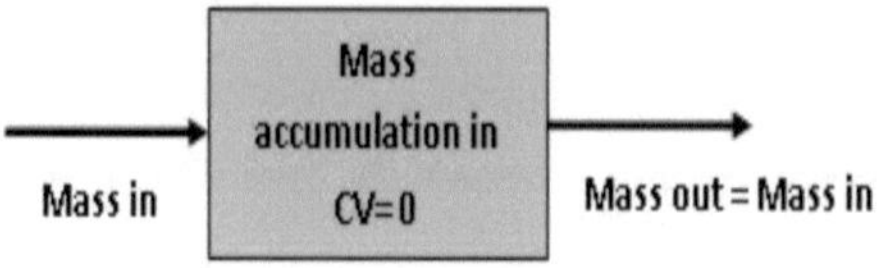

Figure 4.2: *Schematic representation of mass conservation equation in a control volume*

$$Mass\ flow\ in\ CV\ at\ time\ t - mass\ flow\ out\ of\ CV\ at\ time = 0 \quad 4.1.1$$

That means,

$$\frac{\Delta m}{\Delta t} = 0 \quad 4.1.2$$

Considering an infinitesimally small change in time, equation 4.1.2 can be written in its differential form as 4.1.3.

$$\frac{dm}{dt} = 0 \quad 4.1.3$$

Where, *m* is the mass of fluid and *t* is the time.

4.2 Momentum Conservation Equation

The sum of forces on the control volume is given by equation 4.2.1 in the vector form, which is written as equation 4.2.2 in the scalar form.

$$\sum \vec{F} = \vec{F}_P + \vec{F}_g + \vec{F}_f \quad 4.2.1$$

$$\sum F = F_P + F_g - F_f \quad 4.2.2$$

The momentum conservation principle implies that the sum of forces acting on the control volume CV (f_p, f_g and f_f) is balanced by the net momentum flux through the gas in the control volume (Wark and Richards, 1999).

$$\sum F = \frac{d(mu)}{dt} \qquad 4.2.3$$

Where,

F_g is the vertical component of the gravity force

F_p is the pressure force, proportional to the pressure drop or buildup across the control volume.

F_f is the force corresponding to frictions on the wall of the conduct. It acts in the direction opposite to that of flow.

Replacing the terms in equation 4.2.2 by their expressions and putting them in equation 4.2.3and rearranging yields:

$$\frac{d(mu)}{dt} = AdP + \rho g cos\alpha\, Ady - \tau A_s \qquad 4.2.4$$

Where,

τ is the shear stress exerted on pipe segment's wall surface area.

Note that the minus sign in the third term on the right hand side of equation 4.2.4 indicates that the frictional forces act against the direction of flow.

The frictional forces are expressed by equation 4.2.5 in terms of the cross sectional area of the conduit A and its diameter D.

$$F_f = \tau \pi D dy = A \frac{\tau}{A/\pi D} dy = A \frac{4\tau}{D} dy \qquad 4.2.5$$

Where,

$$A = \pi \frac{D^2}{4} \qquad 4.2.6$$

The term on the left hand side of the equation 4.2.4 represents the change of momentum with time. Upon expansion it leads to equation 4.2.7 for a particular conduit's length of constant cross-sectional area.

$$\frac{d(mu)}{dt} = A\Delta y \frac{d(\rho u)}{dt} = \Delta y \frac{d(uA.\rho u)}{dy} \qquad 4.2.7$$

Where,

$$mu = \rho A u \Delta y \qquad 4.2.8$$

$$dt = \frac{dy}{u} \qquad 4.2.9$$

Putting equation 4.2.5 and 4.2.7 into 4.2.4 and rearranging results in equation 4.2.10, which represents the general form of Bernoulli equation, extended by adding frictional term, and valid for an injection well.

$$\frac{dP}{dy} = -\rho g cos\alpha + \rho u \frac{du}{dy} + \frac{4\tau}{D} \qquad 4.2.10$$

The minus sign on the gravitational term in equation 4.2.10, indicates that the hydrostatic pressure contributes to bottom-hole pressure build-up because *dy* has a negative sign. This is the expectation in an injection well.

4.2.1 Frictional Loss Term

The frictional loss in the pipe is a function of the shear stress (τ), which is expressed in terms of fanning friction factor f_f in equation 4.2.11.

$$\tau = \frac{f_f \rho u^2}{2} \qquad 4.2.11$$

The fanning friction factor is a function of the roughness of the surface of the flow conduit and the Reynolds' number (N_{Re}). In a laminar flow regime, where the Reynolds' number is smaller than 2000, the Fanning friction factor is estimated as a simple function of the Reynolds' number represented by equation 4.2.12 (Brill and Mukherjee, 1999, Coulson and Richardson, 1996, Economides et al., 1993).

$$f_f = \frac{16}{N_{Re}} \quad 4.2.12$$

For turbulent and transitional flow regimes with Reynolds' numbers bigger than 2000, the Fanning friction factor can be read from the Moody diagram generated from Colebrook-White equation, represented by equation 4.2.13 below (Brill and Mukherjee, 1999, Coulson and Richardson, 1996, Economides et al., 1993). This equation is also used in this work to calculate Fanning friction factor. Because it is implicit in friction factor, an iterative method is used for its solution.

$$\frac{1}{\sqrt{f_f}} = -4\log\left(\frac{\varepsilon}{3.7065} + \frac{1.2613}{N_{\mathrm{Re}}\sqrt{f_f}}\right) \quad 4.2.13$$

Where,

$$\varepsilon = \frac{e}{D} \quad 4.2.14$$

e is the pipe roughness representing the length of protrusions on the pipe surface available for flow.

The Reynolds' number depends on the gas flow velocity as shown by equation 4.2.15.

$$N_{Re} = \frac{\rho D u}{\mu} \quad 4.2.15$$

Because the product of density and viscosity $(\rho\mu)$ is constant for the flow of compressible fluids (Economides et al., 1993 , P. 145) , the Reynolds' number can also conveniently be expressed by equation 4.2.16.

$$N_{Re} = \frac{4M_g q P_{std}}{\pi D \mu R T_{std}} \quad 4.2.16$$

Where,

M_g is the molecular weight of the gas, q the volumetric flow rate at standard conditions, P_{std} the pressure at standard conditions and T_{std} the temperature at standard conditions and R the universal gas constant.

$$\rho = \frac{M_g P_{std}}{z_{std} R T_{std}} \qquad 4.2.17$$

$$u = \frac{4q}{\pi D^2} \qquad 4.2.18$$

$$z_{std} = 1 \qquad 4.2.19$$

Putting equation 4.2.11, 4.2.17 and 4.2.18 in equation 4.2.5 yields equation which is the developed expression of the frictional forces.

$$F_f = 8 \frac{f_f M_g P_{std} q^2}{\pi D^5 R T_{std}} dy \qquad 4.2.20$$

4.3 Pressure Gradient in a Wellbore

In equations 4.2.10 and 4.2.11, the gas density (ρ) is unknown. It can be determined by the real gas law as a function of molecular weight (M_g), pressure, temperature and compressibility factor (z).

$$\rho = \frac{m}{V} = \frac{M_g P}{zRT} \qquad 4.3.1$$

In equation 4.3.1, all terms on the right hand side are constant except the compressibility factor which is as the density, a function of pressure.

Hence,

$$d\rho = \frac{M_g}{RT} d\left(\frac{P}{z}\right) \qquad 4.3.2$$

Which upon expansion yields

$$d\rho = \frac{M_g}{RT}\left(\frac{zdP - Pdz}{z^2}\right) \qquad 4.3.3$$

However, it is considered that the density and the compressibility factor are constant and assume their average values in each wellbore segment, avoiding the complexity of using equation 4.3.3. In this respect, these parameters refer to their average values in a wellbore segment, in the remaining parts of this work.

Putting equation 4.2.11 and 4.3.1 into 4.2.10 and rearranging gives equation 4.3.4.

$$\frac{dP}{P}=-\frac{M_g}{zRT}gcos\alpha dy+\frac{M_g}{zRT}udu+\frac{2f_fM_gu^2}{DzRT}dy \qquad 4.3.4$$

The mass flow rate ($\dot{m}$) can also be calculated as a function of flow velocity following equation 4.3.5

$$\dot{m}=uA\rho=uA\frac{M_gP}{zRT} \qquad 4.3.5$$

Therefore,

$$u=\frac{\dot{m}zRT}{PAM_g} \qquad 4.3.6$$

Putting equation 4.3.6 into 4.3.4 yields equation 4.3.7.

$$\frac{dP}{P}=-\frac{M_g}{zRT}gcos\alpha dy+\frac{M_g}{zRT}\frac{\dot{m}zRT}{pAM_g}d\frac{\dot{m}zRT}{PAM_g}++\frac{2f_fM_g\left(\dot{m}zRT\right)^2}{DzRT\left(PAM_g\right)^2}dy \qquad 4.3.7$$

Grouping constants and cancelling terms in equation 4.3.7 will yield equation 4.3.8.

$$\frac{dP}{P}=-\frac{M_g}{zRT}gcos\alpha dy+\frac{\dot{m}^2}{A^2M_g}\frac{RT}{P}d\left(\frac{z}{P}\right)+\frac{2f_f\dot{m}^2zRT}{DP^2A^2M_g}dy \qquad 4.3.8$$

Equation 4.3.8 represents the basic pressure equation for single-phase flow of a real gas in a straight conduit in the injection mode. It can be extended to a two-phase flow equation based on the homogenous fluid model. In this sense, the densities as well as other thermodynamic and transport properties of individual phases are combined, weighted by their phase fractions to obtain average single homogenous fluid properties based on equation 4.3.9.

$$X_m=\alpha X_v+\left(1-\chi\right)X_L \qquad 4.3.9$$

Where,

X represents any fluid property

m, *v* and *L* are indices for mixture, vapour and liquid phase respectively

χ is the fraction of the vapour phase

Therefore, equation 4.3.8 will be modified for mixture to yield:

$$\frac{dP}{P}=-\frac{M_m}{z_m RT}g\cos\alpha dy+\frac{\dot{m}^2}{A^2 M_m}\frac{RT}{P}d\left(\frac{z_m}{P}\right)+\frac{2f_{f_m}\dot{m}^2 z_m RT}{DP^2A^2M_m}dy \qquad 4.3.10$$

Equation 4.3.10 will be equivalent to 4.3.8 for a single phase flow of either liquid or gas, in situations where χ =1 or χ =0. Therefore this equation will be used in the remaining parts of this work. It is also worth mentioning that the friction factor of the mixture will be obtained by recombination of phase-dependent properties as described with equation 4.3.9 above.

4.3.1 Dimensional Analysis

Study of problems in fluid dynamics is complicated due to the many parameters which affect them. However quick insight could be obtained by grouping some variables together to form dimensionless groups, with the advantage of reducing effective number of variables (Coulson and Richardson, 1996). Moreover complex problems, characterized by poor understanding of processes and influence of major governing variables can be effectively solved through these dimensionless groups (Brill and Mukherjee, 1999). Coulson and Richardson (1996) further remarked that the basis of dimensional analysis, is the requirement of dimensional consistency, which must be satisfied in terms of fundamental dimensions (mass, length, time and sometimes temperature and moles) and therefore forms a major constraint to dimensional analysis.

The Buckingham Pi Theorem, which states that the number of dimensionless groups is equal to the number of variables minus the number of fundamental dimensions, has been extensively discussed by Coulson and Richardson (1996), who recognized that this theorem provides an easy way of obtaining dimensionless groups during modeling of physical processes. Coulson and Richardson (1996) also pointed out that in some cases, the number of dimensionless groups will be greater than predicted by the Buckingham Pi theorem. Therefore for simple problems, they proposed to express all variables in terms of units of fundamental variables and manipulate them to cancel all dimensions and obtain the dimensionless numbers. This procedure has been applied on equation 4.3.10 to enable derive two important dimensionless numbers for flow analysis in a straight line conduit. In so doing, equation 4.3.10 can be re-written as equation 4.3.11 if the acceleration term is neglected (because of its small contribution of acceleration to pressure change compares with the contributions of gravity and friction in the well) and inlet pressure (P_{in}) and temperature (T_{in}) as well the pipe segment's total length (L) are introduced.

$$\frac{dP}{P} = -f\left(M,l,t,\theta,Mol\right)\frac{T_{in}}{T}\frac{dy}{L} + g\left(M,l,t,\theta,Mol\right)\frac{P_{in}^2}{P^2}\frac{dy}{L} \qquad 4.3.11$$

Where,

M is for mass, t is for time, θ is temperatures, l is for length and Mol is for moles

$$f\left(M,l,t,\theta,Mol\right) - M^1 Mol^{-1} l^1 M^{-1} l^{-2} t^{-2} \theta^1 Mol^1 l^1 t^2 \theta^{-1} M^1 = \dim ensionlessNumber \qquad 4.3.12$$

and

$$g\left(M,l,t,\theta,Mol\right) = M^2 t^{-2} M^1 l^2 t^{-2} Mol^{-1} \theta^{-1} \theta^1 l^1 l^{-1} l^{-4} Mol^1 M^{-1} M^{-2} l^{-2} t^4 l^4 = \dim ensi o n lessnumb \qquad 4.3.13$$

These dimensionless numbers are called N_g and N_f expressed by equation 4.3.14 and 4.3.15 respectively.

N_g represents the ratio of gravitational to pressure forces (Hagoort, 2005). Instead of representing it as introduced by Hagoort for a production well, it has been adapted in this context to relate directly to injection wells. It is equal to the ratio of the average gas' static gradient in the pipe segment to the average pressure in the segment. As such it gives an indication of the percentage of hydrostatic pressure contribution to the total pressure in the well's segment

N_f suggests the measure of the magnitude of frictional forces in the well's segment with respect to pressure forces at the inlet of the segment (Hagoort, 2005). Its magnitude is governed by the conduit diameter, fluid flow-rate and pipe roughness. In this sense, conduit diameter will be chosen so as to minimize frictional losses. However, an injection project is sometimes constrained by existing wellbore configuration, hence tubing diameter and roughness. Therefore, the flow-rate would be the main controlling parameter for the friction number. As a consequence, the absolute value of friction number is often smaller than that of gravity number. At maximum flow rate, it will reach its maximum value. In the absence of gravity forces, say in a horizontal pipe, above this maximum value, flow will no longer be possible.

$$N_g = \frac{M_m L}{z_m R T_{in}} g cos\alpha \qquad 4.3.14$$

$$N_f = \frac{2 z_m f_f \dot{m}^2 R L T_{in}}{D A^2 M_m P_{in}^2} \qquad 4.3.15$$

Introducing equations 4.3.14 and 4.3.15 into equation 4.3.10 and neglecting acceleration term (second term on the right hand side) will yield equation 4.3.16. This equation is used to estimate pressure gradient in an injection well.

$$\frac{dP}{P} = -N_g \frac{T_{in}}{T} \frac{dy}{L} + N_f \frac{P_{in}^2 T}{P^2 T_{in}} \frac{dy}{L} \qquad 4.3.16$$

4.4 Heat Flow in a CO_2 Injection Well

Equations 4.3.10 and 4.3.16 do not include thermal effects. These are the result of the following processes inter alia.

1. Cooling/heating due to gas expansion/compression
2. Heating due to friction
3. Heat exchange with the surrounding environment
4. Heat lost/gained during condensation/evaporation

Modeling of these processes requires application of energy conservation equation modified to suit each situation as a function of the underlying assumptions.

4.4.1 Energy Conservation for a Close System

Derived from the first law of thermodynamics (Wark and Richards, 1999), energy conservation in a close system can be expressed through equation 4.4.1 below.

$$\begin{pmatrix} \text{Change in the} \\ \text{energy of the} \\ \text{system} \end{pmatrix} = \begin{pmatrix} \text{Net transfer of heat} \\ \text{energy into the system} \end{pmatrix} + \begin{pmatrix} \text{Net transfer of energy} \\ \text{into the system by} \\ \text{work interaction} \end{pmatrix} \quad 4.4.1$$

On a unit mass basis, equation 4.4.1 can be mathematically expressed with 4.4.2.

$$dU = \delta Q + \delta W \quad 4.4.2$$

Where,

U is the internal energy of the system, Q is the total heat energy transferred into or out of the system, W is the total net work made on the surrounding by the system or by the surrounding on the system.

The change in work done at constant pressure is obtained in terms of the change in the volume of the system by equation 4.4.3.

$$\delta W = -PdV \quad 4.4.3$$

The enthalpy (H) of a closed system can be defined as a function of its internal energy by equation 4.4.4.

$$H = U + PV \quad 4.4.4$$

Then,

Putting equation 4.4.2 into 4.4.4 and differentiating will yield equation 4.4.5.

$$dH = \delta Q + \delta W + \left(PdV + VdP\right) \quad 4.4.5$$

Upon insertion of equation 4.4.3 in 4.4.5, equation 4.4.6 can be obtained. This equation is the expression of enthalpy as a function of heat transferred in the system or out of the system and the work done on or by the system.

$$dH = dQ + VdP \quad 4.4.6$$

Equation 4.4.6 implies that the change in enthalpy of a closed system corresponds to the maximum amount of energy derivable from a thermodynamic process as pointed out by Wark and Richard (1999).

At constant pressure as assumed at any elevation y in the flow conduit, equation 4.4.6 reduces to 4.4.7, which represents the maximum heat energy transferred in or out of the system.

$$dH = \delta Q \tag{4.4.7}$$

In developing equation 4.4.7 only expansion or compression work was considered. In a wellbore, the work done by the flowing fluid is the sum of expansion/compression work and elevation work (equation 4.4.8)

$$\delta W = -PdV + \delta W_e = -PdV - mg\cos\alpha dy \tag{4.4.8}$$

Putting equation 4.4.8 into 4.4.5 under assumption of constant pressure, and considering an infinitesimally small length dy and letting dH and δQ be the enthalpy and the heat per unit length respectively, equation 4.4.9 can be obtained.

$$dHdy = \delta Qdy - mg\cos\alpha dy \tag{4.4.9}$$

Recalling that the enthalpy is a thermodynamic function depending on pressure and temperature (Wark and Richards, 1999), the total differentiation of enthalpy will give:

$$dH(P,T) = \left(\frac{\partial H}{\partial T}\right)_P dT + \left(\frac{\partial H}{\partial P}\right)_T dP \tag{4.4.10}$$

At constant pressure equation 4.4.10 becomes 4.4.11, expressed in terms of the heat capacity at constant pressure (c_P).

$$dH = \left(\frac{\partial H}{\partial T}\right)_P dT = c_P dT \tag{4.4.11}$$

Therefore,

$$c_P \frac{dT}{dy} = \delta Q - mg\cos\alpha \tag{4.4.12}$$

4.4.2 The Joule-Thomson Effect

It has been established that as the pressure changes in the wellbore the gas will undergo cooling or heating depending on the process and pressure levels (Wark and Richards, 1999). For gases at high pressure, further compression will induce cooling of the gas, whereas at lower pressure, it will induce heating of the gas. Conversely, expansion will induce heating at high pressure and cooling at low pressure. This behavior is known as the Joule-Thomson effect and occurs at constant enthalpy. This isenthalpic process can be performed in the laboratory, where gas is throttled through a valve (see Figure 4.3) or expanded through a porous plate, called diaphragm.

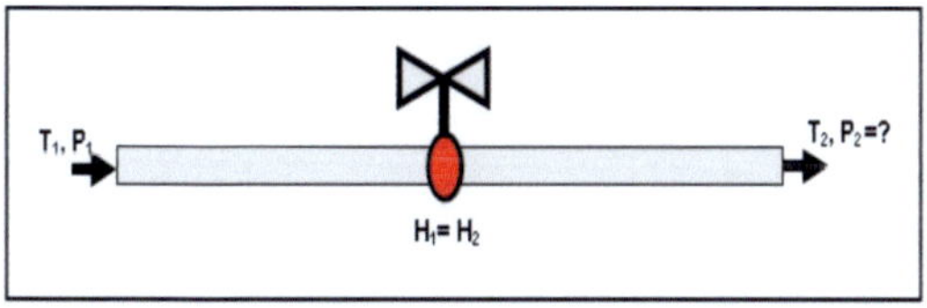

Figure 4.3: *Isenthalpic throttling of gas through throttling valve to observe Joule-Thomson effect*

The relationship between temperature and pressure in a Joule-Thomson experiment is generally represented by Figure 4.4, which shows contrasting behavior of temperature with respect to pressure change, depending on whether the process takes place before or after the inversion temperature as shown on Figure 4.4. Following the constant enthalpy condition (isenthalpic throttling process), equation 4.4.13 representing the gradient of the curve in Figure 4.4, can be used to calculate the Joule-Thomson coefficient (η). The inversion temperature indicates if within the process temperature boundaries, expansion/compression will result in cooling or heating.

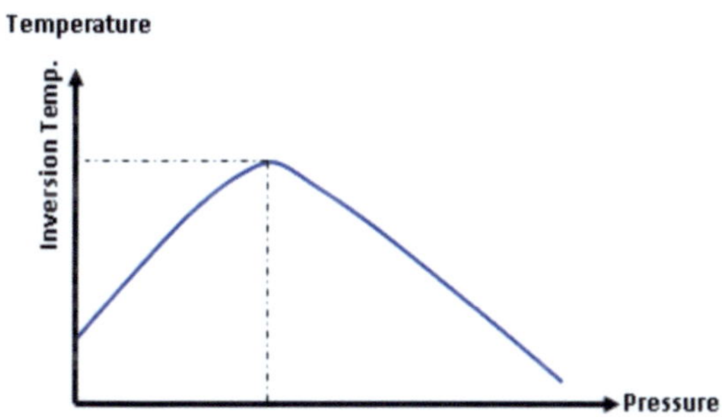

***Figure 4.4**: Temperature behavior during a Joule-Thompson experiment*

$$\eta - \left(\frac{\partial T}{\partial P}\right)_H \qquad 4.4.13$$

From the definition of enthalpy and consequently enthalpy change, equation 4.4.10 will be rewritten as equation 4.4.14 accounting for the constant enthalpy.

$$\left(\frac{\partial H}{\partial T}\right)_P dT + \left(\frac{\partial H}{\partial P}\right)_T dP = 0 \Rightarrow \left(\frac{\partial H}{\partial T}\right)_P dT = -\left(\frac{\partial H}{\partial P}\right)_T dP \qquad 4.4.14$$

Furthermore,

$$\left(\frac{dH}{dP}\right)_T = \left(\frac{\partial H}{\partial T}\right)_P \left(\frac{\partial T}{\partial P}\right)_H = c_p \left(\frac{\partial T}{\partial P}\right)_H \qquad 4.4.15$$

Therefore, putting equation 4.4.15 and 4.4.13 into 4.4.10 yields:

$$\frac{dH}{dy} = c_p \frac{dT}{dy} - c_p \eta \frac{dP}{dy} \qquad 4.4.16$$

Furthermore, putting equation 4.4.16 into 4.4.9 and rearranging, results in the total heat energy gradient per unit mass over an infinitesimally small length (dy) of the flow conduit

$$\delta Q = c_p \frac{dT}{dy} - c_p \eta \frac{dP}{dy} + g\cos\alpha \qquad 4.4.17$$

Equation 4.4.17 can be expressed in terms of molar quantities with respect to equation 4.4.18

$$m = nM_m \qquad 4.4.18$$

Where,

n is the number of moles of the fluid.

Thus for one mole of fluid,

$$\delta q = c_{pm} \frac{dT}{dy} - c_{pm} \eta \frac{dP}{dy} + M_m g\cos\alpha \qquad 4.4.19$$

Where,

c_{pm} is the specific molar heat capacity and M_m is the molecular weight of the fluid.

Rearranging equation 4.4.19, the pressure gradient in the system over an infinitesimally small length dy can be obtained.

$$\frac{dP}{dy} = -\frac{1}{c_{pm}\eta}\delta q + \frac{dT}{\eta dy} + \frac{M_m}{c_{pm}\eta} g\cos\alpha \qquad 4.4.20$$

Putting equation 4.4.20 into 4.3.10 and neglecting kinetic energy term, equation 4.4.21 will be generated.

$$-\frac{\delta q}{c_{pm}\eta}+\frac{dT}{\eta dy}+\frac{M_m}{c_{pm}\eta}g\cos\alpha=-P\frac{M_m}{zRT}g\cos\alpha+P\frac{2f_f M_m(\dot{m}zRT)^2}{DzRT(PAM_m)^2} \quad 4.4.21$$

Rearranging equation 4.4.21 gives equation 4.4.22 that expresses the temperature gradient as a function of the hydrostatic gradient, the frictional gradient and the heat transferred with the surrounding rock.

$$\frac{dT}{dy}=-\frac{M_m}{c_{pm}}g\cos\alpha-\eta P\frac{M_m}{zRT}g\cos\alpha+\eta P\frac{2f_f M_m(\dot{m}zRT)^2}{DzRT(PAM_m)^2}+\frac{\delta q}{c_{pm}} \quad 4.4.22$$

Rearranging equation 4.4.22 yields equation 4.4.23

$$dT=-g\cos\alpha\left(\frac{M_m}{c_{pm}}+\eta P\frac{M_m}{zRT}\right)+\eta P\frac{2f_f M_m(\dot{m}zRT)^2}{DzRT(PAM_m)^2}+\frac{\delta q}{c_{pm}} \quad 4.4.23$$

By introducing equations 4.3.14 and 4.3.15 expressing the dimensionless numbers (N_{gp}) and (N_{fp}), in 4.4.23, equation 4.4.24 can be obtained under assumption that the wellbore transfers heat to the surrounding so that δq is negative.

$$dT=-\left(\frac{zR}{c_{pm}}+\eta\frac{P}{T}\right)N_g T_{in}dy+\left(\eta\frac{T}{P}\right)\frac{P_{in}^2}{T_{in}}N_f dy-\frac{\delta q}{c_{pm}}dy \quad 4.4.24$$

Two additional parameters can be introduced in equation 4.4.24 at this point to yield equation 4.4.25. These parameters have been earlier defined by equation 3.1.3 and 3.1.4. They are expressed in terms of thermodynamic properties of the fluid and therefore they are fluid specific. In this sense, they could also be considered as thermodynamic properties of the fluid stream (Hagoort, 2005).

$$dT=-\varphi N_g T_{in}dy+\omega\frac{P_{in}^2}{T_{in}}N_f dy-\frac{\delta q}{c_{pm}}dy \quad 4.4.25$$

Equations 3.1.3 and 3.1.4 show that Phi and Omega in equation4.4.25 also depend on both temperature and pressure. As a consequence, this equation is non-linear in temperature and its solution requires pressure to be known. The analytical solution can be obtained by using the gas properties at segment's inlet for calculations in the segment. These properties are assumed to be constant in the whole segment. The short length of the segment and the validation criterion[10] imposed on the temperature calculation will preclude sensible errors in the estimates. This assumption is reasonable because all gas properties, apart from omega are weak functions of temperature and pressure, except on the saturation line and in the near-critical region. Furthermore, the fluid temperature is not too sensitive to omega as the latter is a function of the friction number, which is normally very small compared to other parameters.

The segment's boundary conditions for equation 4.4.25 are given by equation 4.4.26.

$$y = 0 : T = T_{in}; y_D = 0 : T_D = 1 \qquad 4.4.26$$

Let dimensionless temperature (T_D) and dimensionless elevation (y_D) be defined by equations 4.4.27 and 4.4.28.

$$T_D = \frac{T}{T_{in}} \qquad 4.4.27$$

$$y_D = \frac{y}{L} \qquad 4.4.28$$

Introducing 4.4.27 and 4.4.28 into 4.4.25 will yield equation 4.4.29 which is the expression for dimensionless temperature per unit length of flow conduit's segment.

$$dT_D = -\varphi N_g dy_D + \omega N_f \frac{P_{in}^2}{T_{in}^2} dy_D - \frac{1}{c_{pm}} \frac{\delta q}{T_{in}} dy_D \qquad 4.4.29$$

A look at equation 4.4.29 shows that φ would control the change in temperature due to gravity forces, and ω, the change in temperature due to friction forces.

Equation 4.4.29 defines the temperature gradient in the flow conduit as a function of gravity effects, frictional effects, heat exchange and other thermodynamic processes. It suggests that the higher φ the higher the temperature increase and conversely, high ω leads to higher temperature fall. It is also noteworthy that φ and ω will cause either heating or cooling depending on the sign of the Joule-

[10] The temperature validation criterion requires that the z-factor at the segment outlet does not differ by more than 0.1% from that at the segment's inlet in absence of phase change. Otherwise the segment length will be chopped until the criterion is met.

Thompson coefficient. However, the above conclusion holds because the injection process will in general take place within the positive Joule-Thomson coefficient region where ω will always induce cooling of the gas and φ induce heating.

5. APPLICATION OF THE MODELS TO WELLBORE INJECTION

This section addresses two instances of the wellbore fluid behavior with respect to thermal interactions with the surrounding environment.

1. Non-adiabatic system behavior
2. Adiabatic closed system behavior (isolated system)

5.1 Non-Adiabatic Flow during Injection

The non-adiabatic situation implies that the fluid interacts thermally with the formation rock, through the wellbore tubular and cement. This results in cooling or heating of the fluid depending on the temperature of the rock and the fluid. Two situations are considered: the simplest one, which assumes that the fluid has enough contact time with the flow conduit's wall, so that its temperature would be considered to have a linear profile and is asymptotical to surrounding rock's temperature; and the more complex one, which assumes that temperature will not be asymptotic to surrounding rock's temperature, but would lie between inlet fluid temperature and surrounding rock's temperature in each segment of the well. Therefore, its estimation will require a rigorous calculation of the overall heat transfer coefficient, which is the main parameter influencing transfer of heat between the formation and the fluid.

5.1.1 Linear Temperature Profile Asymptotic to Formation Temperature

The fluid temperature would be asymptotic to formation temperature in a non-flowing well and whenever the flow rate in a flowing well is small enough for the fluid to have sufficient time to heat up or cool down to a temperature close to that of the surrounding rock.

5.1.1.1 Static Pressure Profile Case

The following simplifying assumptions will be considered in the analysis of non-flowing well:

1. Constant deviation angle in a well segment
2. Constant thermodynamic and transport properties in the segment, taking the segment's average value.

The major consequence of the well not flowing is that the velocity and therefore the mass flow rate are equal to zero. In that respect the friction number will equate to zero and consequently, equation 4.3.16 will become to equation 5.1.1.

$$\frac{dP}{P} = -N_g \frac{T_{in}}{T} \frac{dy}{L} \tag{5.1.1}$$

The temperature gradient corresponding to linear temperature profile can be obtained by equation 5.1.2 as the ratio of the difference between the bottomhole temperature and the wellhead temperature, to the length of the well.

$$\frac{dT}{dy} = \frac{\Delta T}{L} \leftrightarrow dy = \frac{LdT}{\Delta T} \tag{5.1.2}$$

The temperature at any elevation y in the conduit would therefore be expressed in terms of section inlet temperature and temperature gradient as in equation 5.1.3.

$$T = T_{in} + \Delta T \frac{y}{L} \tag{5.1.3}$$

Hence, putting equation 5.1.2 in 5.1.1 yields:

$$\frac{dP}{P} = -N_g \frac{T_{in}}{\Delta T} \frac{dT}{T} \tag{5.1.4}$$

Upon integration of equation 5.1.4 and introducing equation 5.1.3, equation 5.1.5 is obtained.

$$\frac{P_y}{P_{in}} = \left(\frac{T}{T_{in}}\right)^{-\frac{N_g T_{in}}{\Delta T}} = \left(1 + \frac{\Delta T}{T_{in}}\frac{y}{L}\right)^{-\frac{N_g T_{in}}{\Delta T}} \qquad 5.1.5$$

Therefore the pressure at depth y from the conduit's segment's inlet can be obtained by equation 5.1.6

.

$$p_y = p_{in}\left(1 + \frac{\Delta T}{T_{in}}\frac{y}{L}\right)^{-\frac{N_g T_{in}}{\Delta T}} \qquad 5.1.6$$

Whenever $y = L$, the length of conduit segment, the outlet pressure of the conduit's segment is obtained with equation 5.1.7.

$$p_{out} = p_{in}\left(1 + \frac{\Delta T}{T_{in}}\right)^{-\frac{N_g T_{in}}{\Delta T}} \qquad 5.1.7$$

Similarly, the bottomhole pressure could be obtained as a function of the tubing head pressure, using average values of the fluid's properties along the wellbore (equation,5.1.8).

$$p_{bh} = p_{wh}\left(1 + \frac{\Delta T}{T_{wh}}\right)^{-\frac{N_g T_{wh}}{\Delta T}} \qquad 5.1.8$$

The number *e* used in the exponential function can be obtained as a function of a variable *n* when this variable approaches infinity by equation 5.1.9 (Kreyszig, 1999). It is the case if uniform temperature distribution is assumed in equation 5.1.8 , when the temperature is almost constant along the wellbore($\Delta T \rightarrow 0$). In practice this will be seen in isothermal flow, which would be assumed in a horizontal pipe section or in an insulated wellbore where an ideal gas is flowing.

$$\lim_{n\to\infty}\left(1 + \frac{1}{n}\right)^n = e \qquad 5.1.9$$

$$n = \frac{T_x}{\Delta T} \qquad 5.1.10$$

Where T_x stands for section inlet temperature or wellhead temperature.

If equations 5.1.9 and 5.1.10 are put into 5.1.7 and 5.1.8 equations 5.1.11 and 5.1.12 will be obtained respectively.

$$p_{out} = p_{in} e^{-N_g} \quad 5.1.11$$

$$p_{bh} = p_{wh} e^{-N_g} \quad 5.1.12$$

It is worthy to note that assumption of linear temperature distribution, asymptotic to geothermal temperature oversimplifies the problem of a static CO_2 shut-in well. Henninges et al. (2010) and Peterson et al. (Paterson et al., 2008) reported complex temperature distribution in a CO_2 observation well and shut-in wells in general. They identified temperatures higher than formation's temperature in the upper part of the well and lower, somewhere in the middle. In the bottom part of the well, the temperature is described to be close to geothermal temperature. These authors have linked this abnormal temperature distribution to the mass and heat transfer processes occurring in the wellbore, as the liquid falls downward and the vapour rises upward upon condensation and vaporization respectively. This also changes the density distribution in the wellbore and therefore, the pressure distribution. However, there is a tendency for the interplay of all processes that are therein involved, resulting in some compensation, such that the bottomhole pressure does not deviate sensibly from estimated values. Such situation was associated by Paterson et al (2008) to shallow wells or wells in a depleted gas reservoir. This corroborates with Henninges et al. (2010) 's observations, which were made on a well of approximately seven hundred and fifty meters (750 m) deep. Paterson et al (2008) also pointed out that the situation would be different for deeper wells, where a static column can exist in a stable liquid state up to the surface. They further emphasized that the unstable situation would not have a negative effect on injection operations, but would be important for interpretation of reservoir's pressure.

In the light of the foregoing, further work would consist of modeling these mass transfer and heat transfer phenomena with analogy to the phenomena at play in a distillation column.

5.1.1.2 Dynamic Pressure Profile (during injection)

Unlike the previous analysis involving only the gravity pressure effect in the static mode, the dynamic mode incorporates all the three contributors to the pressure behavior in the injection well as anticipated by equation 4.3.10. Consequently, in addition to assumptions 1 and 2 underpinning the non-flowing well case, two other assumptions will be considered.

3. Negligible change in kinetic energy in the well bore segments
4. Constant friction factor in the segment

Based on assumption number 3 and 4 and putting equation 5.1.2 in 4.3.16 and rearranging, equation 5.1.13 can be obtained. This is used as the general equation to predict the pressure profile along a section of the injection well.

$$dP = -N_g \frac{T_x}{\Delta T}\frac{PdT}{T} + N_f \frac{P_x^2}{T_{in}\Delta T}\frac{T}{P}dT \qquad 5.1.13$$

Where, *x* stands for tubing head or section inlet.

Isolating the variables in equation 5.1.13, it can further be represented by equation 5.1.14.

$$dP = K_1 \frac{P}{T}dT + K_2 \frac{T}{P}dT \qquad 5.1.14$$

Where,

$$K_1 = -N_g \frac{T_x}{\Delta T} \qquad 5.1.15$$

$$K_2 = N_f \frac{p_x^2}{T_{in}\Delta T} \qquad 5.1.16$$

A new variable s is used to solve equation 5.1.14 and it is given by equation 5.1.17.

$$s = \frac{P}{T} \qquad 5.1.17$$

Therefore,

$$dP = d(sT) = sdT + Tds \qquad 5.1.18$$

Putting equations 5.1.15 , 5.1.16 and 5.1.18 into 5.1.14 gives equation 5.1.19.

$$sdT + Tds = K_1 sdT + K_2 \frac{1}{s} dT \qquad 5.1.19$$

Separation of variables will yield equation 5.1.20.

$$\frac{dT}{T} = K_3 \frac{sds}{\left(s^2 + K_4\right)} \qquad 5.1.20$$

Where

$$K_3 = \frac{1}{K_1 - 1} \qquad 5.1.21$$

$$K_4 = \frac{K_2}{K_1 - 1} \qquad 5.1.22$$

Upon integration, equation 5.1.20 is obtained.

$$\ln T = K_3 \frac{\ln\left(s^2 + K_4\right)}{2} \qquad 5.1.23$$

By replacing s by its value as presented in equation 5.1.17, equation 5.1.24 can be expressed in terms of pressure and temperature.

$$\ln T = \frac{K_3}{2} \ln\left(\frac{P^2}{T^2} + K_4\right) \qquad 5.1.24$$

Introducing integration boundaries (P_x, T_x) and (P, T) in equation 5.1.24 enables to obtain equation 5.1.25.

$$\ln \frac{T}{T_x} = \frac{K_3}{2} \ln\left[\frac{\left(P^2 / T^2\right) + K_4}{\left(P_x^2 / T_x^2\right) + K_4}\right] \qquad 5.1.25$$

Upon removing the logarithmic terms, equation 5.1.25 results in 5.1.26.

$$\left(\frac{T}{T_x}\right)^{\frac{2}{K_3}} = \frac{\left(P^2/T^2\right) + K_4}{\left(P_x^2/T_x^2\right) + K_4} \qquad 5.1.26$$

Separation of variables leads to equation 5.1.27, which represents the general equation used to build pressure profile with respect to the aforementioned assumptions.

$$P = P_x \sqrt{\left(\frac{T}{T_x}\right)^{2\left(\frac{1+K_3}{K_3}\right)} + K_4 \frac{T^2}{P_x^2}\left[\left(\frac{T}{T_x}\right)^{\frac{2}{K_3}} - 1\right]} \qquad 5.1.27$$

As a function of depth and tubing head conditions and using definition of linear temperature profile equal to surrounding rock's temperature as given by equation 5.1.2, equation 5.1.29 will enable estimate pressure at depth *y* in a pipe section of length *L*.

$$P = P_x \sqrt{\left(\frac{T}{T_x}\right)^{2\left(\frac{1+K_3}{K_3}\right)} + K_4 \frac{T^2}{T_x^2}\frac{T_x^2}{P_x^2}\left[\left(\frac{T}{T_x}\right)^{\frac{2}{K_3}} - 1\right]} \qquad 5.1.28$$

$$P = P_{wh} \sqrt{\left(1 + \frac{\Delta T}{T_{wh}}\frac{y}{L}\right)^{2\left(\frac{1+K_3}{K_3}\right)} + K_4 \left(1 + \frac{\Delta T}{T_{wh}}\frac{y}{L}\right)^2 \frac{T_{wh}^2}{P_{wh}^2}\left[\left(1 + \frac{\Delta T}{T_{wh}}\frac{y}{L}\right)^{\frac{2}{K_3}} - 1\right]} \qquad 5.1.29$$

Therefore the bottomhole pressure and conduit section outlet pressure will be obtained by equations 5.1.30 and 5.1.31 when $L = y$. It is worthy to note that $\frac{y}{L}$ is equation 5.1.29 has negative sign because y is considered negative.

$$P_{bh} = P_{wh} \sqrt{\left(1 - \frac{\Delta T}{T_{wh}}\right)^{2\left(\frac{1+K_3}{K_3}\right)} + K_4 \left(1 - \frac{\Delta T}{T_{wh}}\right)^2 \frac{T_{wh}^2}{P_{wh}^2}\left[\left(1 - \frac{\Delta T}{T_{wh}}\right)^{\frac{2}{K_3}} - 1\right]} \qquad 5.1.30$$

A segment's outlet pressure will similarly be obtained as a function of inlet pressure by replacing variables appropriately in equation 5.1.30 to yield equation 5.1.31.

$$P_{out} = P_{in}\sqrt{\left(1-\frac{\Delta T}{T_{in}}\right)^{2\left(\frac{1+K_3}{K_3}\right)} + K_4\left(1-\frac{\Delta T}{T_{in}}\right)^2 \frac{T_{in}^2}{P_{in}^2}\left[\left(1-\frac{\Delta T}{T_{in}}\right)^{\frac{2}{K_3}} - 1\right]} \quad 5.1.31$$

5.1.2 Temperature Profile Accurately Accounting for Transfer of Heat between Fluid and Surrounding

To arrive at equation 5.1.31, it was assumed that the gas temperature follows a linear profile equal to surrounding rock's temperature as it descends from the well head to the bottom of the hole. It was also assumed that the gas is heated on its way down to the bottomhole which is the case most of the times. However, more realistically, depending on the difference between the gas temperature and the surrounding rock temperature, the result will be a heating up of the gas (whenever the surrounding rock temperature is higher than gas temperature) or cooling of the gas (whenever the gas temperature is higher than the formation temperature). This justifies development of a representative temperature profile equation (see equation 4.4.29), which will be substituted into equation 5.1.31 to yield the pressure solution for a straight flow conduit of constant diameter. Therefore, introducing equations 4.4.29 into 5.1.31, mindful the definitions of dimensionless temperature and depth, given by equations 4.4.27 and 4.4.28 respectively, equation 5.1.32 can be obtained. This equation enables estimation of pressures along the wellbore, provided that the temperatures are known. Therefore calculations would involve two stages at each wellbore's segment: (i) solving for temperature using equation 4.4.29, and (ii) solving for pressure with equation 5.1.32.

$$P_{out} = P_{in}\sqrt{\left(1-\Delta T_D\right)^{2\left(\frac{1+K_3}{K_3}\right)} + K_4\left(1-\Delta T_D\right)^2 \frac{1}{P_{in}^2}\left[\left(1-\Delta T_D\right)^{\frac{2}{K_3}} - 1\right]} \quad 5.1.32$$

Solution of equation 4.4.29 requires knowledge of the amount of heat transferred between fluid and surrounding rock, which implies knowledge of the overall heat transfer coefficient between the fluid and the rock-cement interface as well as the time function for the propagation of heat in the surrounding rock. The analysis that follows concentrates on these important parameters. They are estimated based on the heat energy conservation equations and the basic assumptions considered by earlier workers to analytically estimate heating in a steam injection well and in a gas production wells (Hagoort, 2004,

Hagoort, 2005, Ramey Jr., 1962, Satter, 1965, Willhite, 1967, Wu and Pruess, 1990). These are as follows:

1. Temperature of the fluid is a function of depth *Z* and time *t* meaning *T(Z,t)*.
2. Outer temperature of the casing is also function of *Z* and *t*
3. Heat transmission in the wellbore is rapid as compared to heat flow in the formation and thus could be considered as steady-state.
4. Heat transferred in the formation is by unsteady radial conduction
5. Physical and thermal properties of earth and those of wellbore fluid do not vary significantly with temperature and depth.
6. Gas film coefficients and thermal resistance of annular fluid or insulating materials often have the greatest effect on the overall heat transfer coefficient
7. Fluid flow in the tubing is 1D, steady and with constant mass flow rate
8. The vertical heat conduction in the fluid is neglected compared to heat convection by flowing fluid.
9. In the surrounding rock, the initial geothermal gradient is a known linear function of depth.
10. The vertical heat conduction in the formation can be ignored compared to radial heat conduction

Inclusion of heat uptake/loss in the energy conservation equation is based on the approach of Ramey (1962), which as mentioned in the above assumptions 3 and 4, considers that heat loss/gained from/by the wellbore to the surrounding formation or vice-versa, is determined by steady-state transfer through the wellbore components and transient heat transfer by radial conduction in the formation. As a consequence it will actually result in pseudo-steady-state transfer through the wellbore components.

The wellbore tubing and/or casing through which the gas flows are considered as buried pipes so that the convective heat exchange with the surrounding can be neglected. Therefore, analysis focuses on radial conductive, convective and irradiative heat transfer in the wellbore.

The heat energy conservation is represented by equation 5.1.33.

$$\begin{pmatrix} Rate\,of \\ heat\,in \end{pmatrix} - \begin{pmatrix} Rate\,of \\ heat\,out \end{pmatrix} + \begin{pmatrix} Rate\,of \\ heat\,generation \end{pmatrix} = \begin{pmatrix} Rate\,of\,heat \\ accumulation \end{pmatrix} \qquad 5.1.33$$

Under steady state, the accumulation term is equal to zero except in the gas stream or in the formation, depending on whether these are heat sink or heat source at the depth of interest. It is also considered that no heat is generated in the system.

Thus, for all well components from the formation to the gas stream, equation 5.1.33 will be reduced to equation 5.1.34.

$$\begin{pmatrix} Rate\,of \\ heat\,in \end{pmatrix} = \begin{pmatrix} Rate\,of \\ heat\,out \end{pmatrix} \quad 5.1.34$$

5.1.2.1 Estimation of the Overall Heat Transfer Coefficient

The rate of the steady-state heat transferred from the wellbore to the surrounding rock and vice-versa is proportional to the temperature difference between the fluid and the surrounding rock and to the heat transfer area, the surface area perpendicular to the direction of heat flow. Willhite (1967) recognized that the proportionality factor is the overall heat transfer coefficient. It represents also the net resistance of the tubing, casing annulus, casing wall and cement sheath to heat flow. He proposed an expression for the estimation of the overall heat transfer coefficient. This expression is a function of the heat transfer area. For simplification he considered the surface area as a characteristic area which could be the outer surface area of the tubing in case of injection through the tubing, or the inside surface area of the casing in case of injection through the annulus. Knowing that the rate of heat transfer is proportional to the surface area, this assumption of characteristic area may underestimate the area available for the transfer of heat because, for instance the outer tubing surface area is much smaller that the surface area of the cement sheath. As such it might lead to errors in the estimates.

In order to minimize these errors, this work estimates the overall heat transfer coefficient as a function of the overall heat transfer area. This area is calculated as the log-average of the smallest and largest surface areas in the wellbore (inside tubing and outside and rock-cement interface area in case of injection through tubing). It is assumed that the wellbore is a composite wall made up of cement sheath, casing wall, casing annulus and tubing. Coulson and Richardson (1996, Pp. 383-393) in discussing heat transfer through thick-walled material made the analogy with composite materials. Their methodology has been adopted in this work. However modifications are performed to account for the fact that the heat transfer processes in the annulus and between the tubing surface and the flowing fluid are different from the heat transfer processes through the tubing, the casing and the cement sheath.

Three major heat transfer processes have been identified for a wellbore system under assumption of steady-state heat flow (Willhite, 1967). These are conduction, through the solid components (cement, casing and tubing or annulus filled with insulating materials); convection (natural for stationary fluids and forced for a flowing fluid) and radiation in the annulus filled with a fluid. Figure 5.1 gives a schematic representation of the composite wall model and the transfer coefficients that govern heat transfer in each component.

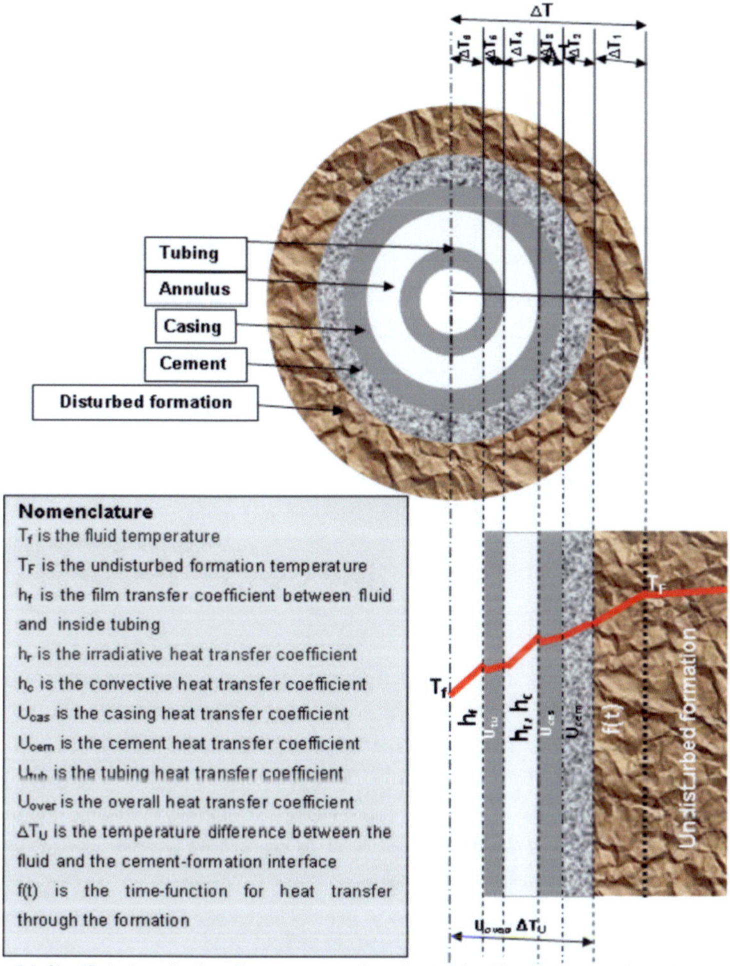

Figure 5.1: *Simplified schematic of a wellbore as a composite wall material, guiding development of overall heat transfer coefficient in analogy with a thick-walled tube model*

5.1.2.1.1 Heat Transfer by Steady-state Conduction

Coulson and Richardson (1996) recalled the equation for calculation of the rate heat flow in radial coordinates (equation 5.1.35).

$$Q = -k2\pi rl\frac{dT}{dr} \quad 5.1.35$$

Where,

Q is the rate of heat transfer, l is the length of the pipe, k is the thermal conductivity of the pipe's material.

Integrating equation 5.1.35 between the inner (r_{ii}) and outer (r_{io}) radii of a wellbore component, will yield equation 5.1.36.

$$Q = \frac{2\pi lk\Delta T}{\ln\left({r_{io}}/{r_{ii}} \right)} \quad 5.1.36$$

Equation 5.1.36 can be expressed as a function of the logarithmic average (r_{av}) of the inner and outer radius of a layer, as in equation 5.1.37.

$$Q = \frac{k2\pi r_{av} l\Delta T}{r_o - r_i} \quad 5.1.37$$

Where,

$$r_{av} = \frac{r_o - r_i}{\ln\left({r_{io}}/{r_{ii}} \right)} \quad 5.1.38$$

The rate of heat transferred by conduction is calculated by Fourier's law (Willhite, 1967, Coulson and Richardson, 1996), given in equation 5.1.39 below.

$$Q = -UA_s(\Delta T) \quad 5.1.39$$

Where,

A_s is the area available for heat transfer and U is the heat transfer coefficient.

A can be estimated for a particular layer and as a function of the logarithmic mean average of the inner and outer radii of a pipe, with equation 5.1.40.

$$A_s = 2\pi r_{av} l \quad 5.1.40$$

Comparing equations 5.1.39 and 5.1.36 and 5.1.37 enables to isolate the heat transfer coefficient (U) from equation 5.1.36. This is expressed by equation 5.1.41.

$$U_i = \frac{k_i}{r_{avi} \ln\left(r_{io} / r_{ii} \right)} \quad 5.1.41$$

l is the length of the pipe segment and r_{av} is the logarithmic mean radius of the inner and the outer radius of the layer.

As shown in figure 5.1, the wellbore system is a composite wall made up of n components with different heat transfer coefficients and also different surface areas available for heat flow. The temperature difference between neighboring layers (component of the wellbore) is obtained by rearranging equation 5.1.39 to obtain 5.1.42 expressed in terms of individual layers ***i***.

$$\Delta T_i = -\frac{Q}{A_{si}} \frac{1}{U_i} \quad 5.1.42$$

5.1.2.1.2 Heat Transfer by Natural Convection in the Annulus

Heat transfer in the annulus filled with a fluid will take place by natural convection as earlier-mentioned. The density of the heated fluid next the tubing/casing wall is less than the cooler one next casing/tubing. The measure of the density difference, which creates the buoyancy force and hence the fluid convection, is the product of thermal expansion coefficient and temperature difference (Coulson and Richardson, 1996).

Hasan and Kabir (1994) and Willhite (1967) proposed equations for natural convective heat transfer coefficient between concentric surfaces (equation 5.1.43 to 5.1.45 below). These are based on two dimensionless numbers: the Prandtl (N_{Pr}) and the Grashof (N_{Gr}) numbers.

$$h_{an} = \frac{0.049\left(N_{Gr} N_{Pr}\right)^{1/3} N_{Pr}^{0.074} K_{an}}{r_{to} ln\left(\frac{r_{ci}}{r_{to}}\right)} \qquad 5.1.43$$

Where,

h_{an} is the hantural convection heat transfer coeffeicient in the annulus and K_{an} K_{an} is the thermal conductivity of the annulus fluid

The Grashof number indicates the extent of motion of the annulus fluid as a consequence of natural convection, while the Prandtl number provides a measure of the interaction between hydrodynamic boundary layers and thermal boundary layers (Brill and Mukherjee, 1999, Coulson and Richardson, 1996). They are defined by equations 5.1.44 and 5.1.45 respectively.

$$N_{Gr} = \frac{\left(r_{ci} - r_{to}\right)^3 g \rho_{an}^2 \beta \left(T_{to} - T_{ci}\right)}{\mu_{an}^2} \qquad 5.1.44$$

$$N_{Pr} = \frac{\mu_{an} C_{P_{an}}}{K_{an}} \qquad 5.1.45$$

Where,

μ_{an}, ρ_{an}, C_{pan} and β are viscosity, density, heat capacity and thermal expansion coefficient of annular fluid respectively and g is the acceleration of gravity.

5.1.2.1.3 Heat Transfer by Radiation through the Annulus

Irradiative heat transfer does not require presence of a fluid or a solid to occur. It can also be present in a vacuum as well as being effected through a fluid. It will occur whenever a straight line can be drawn between two surfaces, that is, the surfaces are facing each other. Another prerequisite for radiation transfer is that the facing surfaces should have temperatures above zero Kelvin (Coulson and Richardson, 1996). The irradiative heat flux is calculated based on the Stefan-Boltzmann's law, adapted to wellbore annulus through equation 5.1.46 below (Coulson and Richardson, 1996, Willhite, 1967).

$$Q_r = 2\pi r_{to}\sigma F_{tci}\left(T_{to}^{*4} - T_{ci}^{*4}\right)\Delta L \qquad 5.1.46$$

Where,

σ is the Stefan-Boltzmann constant given by equation 5.1.47.

$$\sigma = 5.67 * 10^{-8} Wm^{-2}K^{-4} \qquad 5.1.47$$

T^* is the absolute temperature and F_{tci} is the view factor, which corresponds to the fraction of radiation emitted from external surface area of tubing or internal surface area of casing to internal surface area of casing or external surface area of tubing, the hotter surface emitting to the colder one. F_{tci} is calculated with equation 5.1.48 below.

$$\frac{1}{F_{tci}} = \frac{1}{I_{tci}} + \left(\frac{1}{E_{to}} - 1\right) + \frac{A_{sto}}{A_{sci}}\left(\frac{1}{E_{ci}} - 1\right) \qquad 5.1.48$$

Where,

E is emissivity, A are surface areas and I_{tci} is the overall heat interchange factor between the two surfaces. Willhite (1967) recommended the value 1.0 for I_{tci}.

In light of the foregoing equation 5.1.46 could be rewritten as 5.1.49, which is a function of temperature difference between the two surfaces.

$$Q_r = 2\pi r_{to}h_r\left(T_{to}^{*} - T_{ci}^{*}\right)\Delta L \qquad 5.1.49$$

Where h_r is the radiation heat transfer coefficient defined by equation 5.1.50.

$$h_r = \sigma F_{tci}\left(T_{to}^{*2} + T_{ci}^{*2}\right)\left(T_{to}^{*} + T_{ci}^{*}\right) \qquad 5.1.50$$

5.1.2.1.4 The Fluid's Film Transfer Coefficient

The heat transfer from the inner surface of the flow conduit to the fluid and vice-versa is dominated by convection, especially in the presence of turbulence. This is because fluid in general and CO_2 in particular are very pour heat conductors. In this work, the effect of natural convection is neglected and the heat flow is treated only with respect to forced convection inside a tube.

The basis to compute the heat transfer coefficient from the flow conduit's inner surface to the flowing gas is the equation proposed by Colburn (Coulson and Richardson, 1996). This equation represents heat flow as a function of the j-factor (equation 5.1.51) for heat transfer in terms of a set of dimensionless numbers, which are specified below.

$$J_H = N_{St} N_{Pr}^{\;0.67} = 0.023 N_{Re}^{\;-0.2} \qquad 5.1.51$$

- The Reynold's number:

$$N_{Re} = \frac{uD\rho}{\mu} \qquad 5.1.52$$

- The Nusselt number

$$N_{Nu} = \frac{h_f D}{k_f} \qquad 5.1.53$$

- The Prandtl number:

$$N_{Pr} = \frac{c_p \mu}{k_f} \qquad 5.1.54$$

- The Stanton number

$$N_{St} = N_{Nu} N_{Re}^{\;-1} N_{Pr}^{\;-1} = \frac{J_H}{N_{pr}^{\;0.67}} = \frac{h_f}{c_p \rho u} \qquad 5.1.55$$

Then, by rearranging equation 5.1.51, another expression of the Stanton number could be obtained as in equation 5.1.56.

$$N_{St} = 0.023 N_{Re}^{\;-0.2} N_{Pr}^{\;-0.67} \qquad 5.1.56$$

Where for all the equation above,

u is the fluid velocity, D the flow conduit's diameter, μ the fluid viscosity, k_f the fluid thermal conductivity, h_f the fluid's film transfer coefficient and ρ is the fluid's density.

All physical properties are normally determined at the bulk temperature (mean fluid's temperature at the center of the flow conduit), apart from the viscosity, in the Reynolds' number, which is evaluated at the mean film temperature ($\overline{T}_f$), directly near the conduit's surface.

$$\overline{T}_f = \frac{T_{s+}T_{bulk}}{2} \qquad 5.1.57$$

However, because the viscosity of CO_2 is normally low and has a relatively little sensitivity to temperature, of which difference between bulk fluid and pipe surface will not be large in the wellbore, the viscosity in the Reynolds' number can also be estimated at the bulk fluid temperature.

Upon rearranging equation 5.1.55, the film transfer coefficient can be determined from the Stanton number.

$$h_f = N_{St}\rho c_p u \qquad 5.1.58$$

At this point where, all the different expressions for heat transfer coefficients in individual components of the wellbore have been determined, the calculation of the overall heat transfer coefficient from the formation-cement interface to the tubing inner surface can be performed. The overall temperature drop is the sum of the temperature drops in the different components, which is obtained by applying equation 5.1.42 to all the wellbore components and adding them together.

$$\Delta T_{over} = Q\frac{1}{A_{sover}U_{over}} = \Delta T_1 + \Delta T_2 + \ldots + \Delta T_n = Q\left(\frac{1}{A_{s1}U_1} + \frac{1}{A_{s2}U_2} + \ldots + \frac{1}{A_{sn}U_n}\right) \qquad 5.1.59$$

By replacing U_i in 5.1.59 by their corresponding expressions for all components in the wellbore, 5.1.60 is obtained.

$$T_{ti} - T_{fc} = Q\left(\frac{1}{A_{sti}h_f} + \frac{1}{\overline{A}_{stub}U_{tub}} + \frac{1}{\overline{A}_{san}\left(h_r + h_c\right)} + \frac{1}{\overline{A}_{scas}U_{cas}} + \ldots + \frac{1}{\overline{A}_{scem}U_{cem}}\right) \qquad 5.1.60$$

Therefore, the overall heat transfer coefficient from the formation-cement interface through wellbore components up to the flow conduit's inside surface can be estimated by equation 5.1.61, obtained by mapping equation 5.1.60 to equation 5.1.59 and rearranging the variables

$$\frac{1}{U_{over}} = A_{sover}\left(\frac{1}{A_{sti}h_f} + \frac{1}{\overline{A}_{stub}U_{tub}} + \frac{1}{\overline{A}_{san}\left(h_r + h_c\right)} + \frac{1}{\overline{A}_{scas}U_{cas}} + \ldots + \frac{1}{\overline{A}_{scem}U_{cem}}\right) \qquad 5.1.61$$

For radial flow of heat, the surface area available for heat flow can be calculated by equation 5.1.62.

$$A_s = 2\pi r l \tag{5.1.62}$$

Using equation 5.1.62 to estimate the overall heat transfer area as a function of the logarithmic average radii of the first and last layer (r_{av1} and r_{avn}) and that of individual layers as a function of logarithmic average of their inner and outer radii, yields equation 5.1.63 and 5.1.64

$$A_{sover} = 2\pi \left[\frac{r_{avn} - r_{av1}}{\ln\left(r_{avn} / r_{av1} \right)} \right] l \tag{5.1.63}$$

$$\overline{A}_{si} = 2\pi \left[\frac{r_{io} - r_{ii}}{\ln\left(r_{io} / r_{ii} \right)} \right] l \tag{5.1.64}$$

Putting equations 5.1.63 and 5.1.64 into 5.1.61 in light of equation 5.1.62, results in equation 5.1.65, which is the expression used to calculate overall heat transfer coefficient in this work.

$$\frac{1}{U_{over}} = \frac{r_{over}}{r_{tubin} h_f} + \frac{r_{over}}{r_{avtub} U_{tub}} + \frac{r_{over}}{r_{avan}\left(h_c + h_r\right)} + \frac{r_{over}}{r_{avcas} U_{cas}} + \ldots + \frac{r_{over}}{r_{avcem} U_{cem}} \tag{5.1.65}$$

Where,

$$r_{over} = \frac{r_{avn} - r_{av1}}{\ln\left(r_{avn} / r_{av1} \right)} \tag{5.1.66}$$

$$r_{avi} = \frac{r_{io} - r_{ii}}{\ln\left(r_{io} / r_{ii} \right)} \tag{5.1.67}$$

r_{tubin} is the tubing inner radius, r_{avtub} is the average radius of tubing, r_{avan} is average radius of annulus, r_{avcas} is average radius of casing and r_{avcem} is average radius of cement sheet. And indices tub, cas, an and cem stand for tubing, annulus, casing and cement respectively.

5.1.2.1.5 Temperature of the Fluid in the Flow Conduit

The rate of heat transfer between the surrounding rock and the fluid or vice-versa is obtained by applying the heat energy conservation equation with respect to the steady-state heat transfer condition (equation 5.1.68).

$$U_{over}A_{over}\left(T_{fc}-T_{in}\right)=c_p\rho q\left(T_{out}-T_{in}\right) \qquad 5.1.68$$

Where,

T_{fc} is the formation-cement interface temperature and T_{cin} is the conduit inside temperature, T_{out} is the fluid's final temperature at the segment's outlet and T_{in} the fluid's initial temperature at the segment's inlet. Equation 5.1.68 considers the influence of heat exchange with the surrounding only and neglects other heating processes.

Rearranging equation 5.1.68 to calculate T_{out} will give equation 5.1.69.

$$T_{out}=\frac{U_{over}A_{over}\left(T_{fc}-T_{in}\right)}{c_p\rho q}+T_{in} \qquad 5.1.69$$

Based on dimensional analysis, a new dimensionless number can be established through equation 5.1.70.

$$\aleph=\frac{U_{over}A_{over}}{c_p\rho q}=\frac{U_{over}A_{over}}{c_p\dot{m}} \qquad 5.1.70$$

Therefore considering only heat transferred between the fluid and the surrounding rock,

$$T_{out}=\aleph\Delta T_{fF}+T_{in} \qquad 5.1.71$$

Where, T_{out} is the segment outlet temperature and T_{in} is the segment inlet fluid temperature

$$\Delta T_{fF}=T_{fe}-T_{in} \qquad 5.1.72$$

T_{fc} is the formation-cement interface temperature at the depth of the segment's inlet, which is a function of undisturbed rock geothermal gradient, elevation and time function for heat conduction (equation 5.1.73); and T_{in} is the fluid's temperature at the inlet of the segment.

$\aleph$ represents the ratio of the heat transfer capacity between the formation and the fluid, to the fluid heat capacity. As emphasized by Hagoort (2005), it is the measure of heat lost to or gained from the formation with respect the heat transported through the flow conduit assuming that all heat transported is absorbed by the fluid under steady state transfer. Consequently, high values would indicate lower resistance of the wellbore components to heat flow and/or low fluid's heat capacity. Its value also depends on the fluid's flow rate. At high flow rates, there will be little contact time between the gas and the flow conduit's inner surface and consequently low heating interactions. Therefore the value of $\aleph$ would be smaller and as a consequence the flow will approach adiabatic conditions. Conversely, lower flow rates would suggest higher values of $\aleph$ and the influence of the surrounding rock temperature will be more pronounced.

5.1.2.1.6 Temperature at the Formation-Cement Interface

As earlier emphasized, heat conduction through the surrounding rock is radial and transient. Brill and Mukherjee (1999) recalled that the transient-radial heat conduction equation is identical to the diffusivity equation used in transient well test analysis to calculate reservoir's pressure response, under assumption that the well is a linear line source in an infinite reservoir. Therefore, they proposed the line source solution, earlier introduced by Ramey (1962), to calculate the formation-cement interface temperature. This is expressed by equation 5.1.73.

$$T_{fc} = T_F + \frac{q}{2\pi\Delta L}\frac{f(t)}{K_f} \qquad 5.1.73$$

Where,

T_F is the undisturbed rock temperature (equation 5.1.75), q the heat flow rate, K_f the rock's thermal conductivity and $f(t)$ the transient heat conduction time function, also known as the Ramey's time function and given by equation 5.1.74.

$$f(t) = \frac{1}{2}E_i\left(\frac{r_w^2}{4\alpha t}\right) \qquad 5.1.74$$

$$T_F = y\Delta T_g \qquad 5.1.75$$

Where,

r_w is the wellbore radius; *t* the injection duration; E_i the exponential integral; α is the surrounding rock's thermal diffusivity and y is the depth of the well segment . α is be expressed by equations 5.1.77.

$$\Delta T_g = \frac{T_{res} - T_s}{y} \tag{5.1.76}$$

T_{res} is the reservoir temperature, T_s is surface temperature and y is the depth of the well.

$$\alpha = \frac{K_f}{\rho_f C_f} \tag{5.1.77}$$

ρ_f and C_f are surrounding rock's density and heat capacity respectively.

For injection times greater than one week when $\frac{r_w^2}{4\alpha t} < 0.0025$, E_i's solution can be approximated by a logarithmic function as in equation 5.1.78 and $f(t)$ by equation 5.1.79.(Brill and Mukherjee, 1999)

$$E_i\left(-\frac{r_w^2}{4\alpha t}\right) = \ln\left(-\frac{r_w^2}{4\alpha t}\right) + 0.5772 \tag{5.1.78}$$

$$f(t) = 0.405 + 0.5\ln(t_D) \tag{5.1.79}$$

Where, t_D is the dimensionless time and is given by equation 5.1.80 below.

$$t_D = \frac{\alpha t}{r_w^2} \tag{5.1.80}$$

The validity of $f(t)$ as expressed by equation 5.1.74 is restricted to time intervals above one week, corresponding to 168 hours. Hassan and Kabir (1991) further proved that equation 5.1.74 will result to large errors if applied to time less than 250 hours. They went on to propose two equations corresponding to early and late time respectively (equations 5.1.81 and 5.1.82). These equations when used together are valid for all times.

For $t_D \le 1.5$

$$f(t) = 1.1281\sqrt{t_D}\left(1 - 0.3\sqrt{t_D}\right) \tag{5.1.81}$$

For $t_D > 1.5$

$$\Delta T_{fF} = T_{fc} - T_{in} \tag{5.1.82}$$

Still in a bid of improving temperature estimation at early time, Eppelbaum and Kutasov (2006) recognized that considering the well as an infinite cylindrical source rather than as a line source would enable to minimize errors in the estimation of the time function at times smaller than one week. They pointed out that for practical purposes, a cylinder whose length is five times or more its diameter could be treated as an infinite cylinder. This is the case for wellbores and wellbore segments considered in this work. Consequently, the principle of superposition will apply at all times be it early or late. They went on to propose an expression to calculate the wellbore wall temperature which corresponds to the rock-cement interface temperature in this context. This expression is given by equation 5.1.83 below.

$$T_{fc} = T_F + \frac{q}{2\pi K_f \Delta L} \ln\left[1 + \left(c - \frac{1}{a + \sqrt{t_D}}\right)\sqrt{t_D}\right] \tag{5.1.83}$$

Where,

$$a = 2.701505 \tag{5.1.84}$$

$$c = 1.4986055 \tag{5.1.85}$$

By comparing equation 5.1.83 with equation 5.1.73 an expression equivalent to the Ramey's time function can be isolated from equation 5.1.83. This expression is the time function that is used in the work to predict the propagation of temperature in the rock surrounding the wellbore. It is given by equation 5.1.86.

$$f(t) = \ln\left[1 + \left(c - \frac{1}{a + \sqrt{t_D}}\right)\sqrt{t_D}\right] \tag{5.1.86}$$

The validity of this time function for all injection times has been verified on figure 5.2 below. This validity confirms the fact that the principle of superstition would hold without limitations. Therefore the model can be confidently used to also estimate early-time temperatures in the wellbore and the surrounding rock.

Figure 5.2 shows estimations of time function with equations of Ramey, Hassan and Kabir and Eppelbaum and Kutasov respectively. It reveals that Ramey's solution will not be valid for t_D<1.5 as expected. After this point of time, the three solutions get closer and converge for longer time periods. Eppelbaum and Kutasov's solution aligns earlier on Ramey's own. Therefore it is the one used as first choice in this work, although Hassan and Kabir's solution is implemented as a check tool.

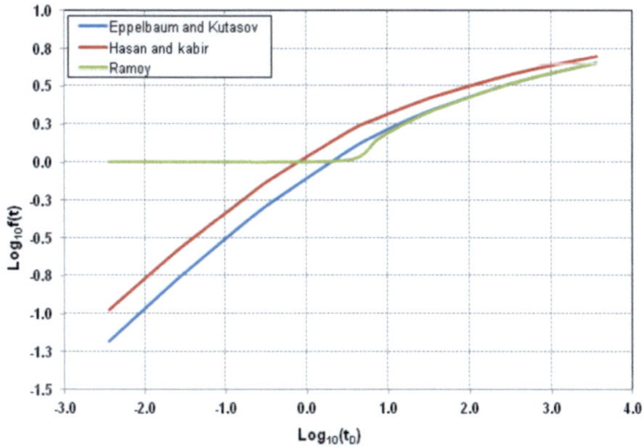

***Figure 5.2**: Comparison of time function solutions and their time-range validity*

At this point the formation-cement interface temperature can be determined by fixing equation 5.1.86 into equation 5.1.73 to give equation 5.1.87. Further derivation will be provided in the section dealing on the estimation of flow conduit inner temperature.

$$T_{fc} = T_F + \frac{q}{2\pi\Delta L} \frac{\ln\left[1+\left(c-\frac{1}{a+\sqrt{t_D}}\right)\sqrt{t_D}\right]}{K_f} \quad 5.1.87$$

The fluid temperature in the flow conduct segments (under the influence of heat transfer between fluid and surrounding rock only) could be expressed by putting equation 5.1.72 in 5.1.71 and rearranging to yield equation 5.1.88.

$$T_{out} = \aleph\left(T_{fc} - T_{in}\right) + T_{in} = \aleph T_{fc} + T_{in}\left(1-\aleph\right) \quad 5.1.88$$

The final gas temperature can be obtained knowing initial temperature in a well segment, by rearranging equation 5.1.88 to give 5.1.89.

$$T_{out} = T_{in} + \aleph\left(T_{fc} - T_{in}\right) \quad 5.1.89$$

The rate of heat absorbed by the fluid per unit length is expressed by equation 5.1.90.

$$\dot{Q} = c_p \rho q \left(T_{out} - T_{in}\right) \qquad 5.1.90$$

Where,

$$T_{out} - T_{in} = \aleph\left(T_{fc} - T_{in}\right) \qquad 5.1.91$$

And is obtained by rearranging equation 5.1.89.

$\dot{Q}$ can also be calculated in terms of formation-cement interface temperature by equation 5.1.92, obtained by introducing equation 5.1.91 and rearranging.

$$\dot{Q} = c_p \rho q \aleph\left(T_{fc} - T_{in}\right) \qquad 5.1.92$$

Upon further rearrangements and recalling definition of mass flow rate as the product of the density and the volumetric flow rate, equation 5.1.93 is obtained.

$$\dot{Q} = c_p \dot{m} \aleph\left(T_{fc} - T_{in}\right) = c_p \dot{m} \aleph \Delta T_{fF} \qquad 5.1.93$$

The heat flow rate can also be estimated in terms of molar quantities by equation 5.1.94.

$$\dot{q} = c_{pm} \frac{\dot{m}}{M_m} \aleph \Delta T_{fF} \qquad 5.1.94$$

Where,

$\dot{q}$ is the heat flow rate per mole of gas and c_{pm} is the molar specific heat capacity at constant pressure.

It is worthy to note that the temperature difference between the cement-formation interface and the fluid (ΔT_{fF}) is the driving force to maintain a constant heat transfer rate from the surrounding rock to wellbore fluid and vice versa. The heat transferred to the fluid as well as the temperature difference will vary as a function of depth (y) along the well. However it will be constant for each depth and time.

Therefore equation 5.1.94 could be re-written to account for the change in heat flow rate as a function of well depth, as equation 5.1.95.

$$\Delta \dot{q} = c_{pm} \frac{\dot{m}}{M_m} \aleph \frac{\Delta T_{fF}}{\Delta y} \tag{5.1.95}$$

For an infinitesimally small segment length, equation 5.1.95 becomes:

$$d\dot{q} - c_{pm} \frac{\dot{m}}{M_m} \aleph \frac{dT_{fF}}{dy} \tag{5.1.96}$$

Therefore the total amount of heat that will flow between the rock and the fluid in a particular segment per unit mole will be expressed by equation 5.1.97. Cancelling terms from this equation yields equation 5.1.98.

$$dq_h = c_{pm} \frac{\dot{m}}{M_m} \aleph \frac{dT_{fF}}{dy} \frac{M_m}{\dot{m}} \tag{5.1.97}$$

$$dq_h = c_{pm} \aleph \frac{dT_{fF}}{dy} \tag{5.1.98}$$

Multiplying both sides of equation 5.1.98 by dy and substituting it into equation 4.4.29 will give 5.1.99.

$$dT_D = -\varphi N_g dy_D + \omega N_f \frac{P_{in}^2}{T_{in}^2} dy_D - \aleph \frac{dT_{fF}}{T_{in}} dy_D \tag{5.1.99}$$

Where dT_{fF} is the temperature difference between the formation-cement interface and the fluid.

Integrating equation 5.1.99 between the inlet and the outlet of a flow conduit's segment equation 5.1.100 is obtained in terms of dimensionless temperatures.

$$T_{Dout} - T_{Din} = -\varphi N_g \left(y_{Dout} - y_{Din}\right) + N_f \omega \frac{P_{in}^2}{T_{in}^2} \left(y_{Dout} - y_{Din}\right) - \aleph \left(T_{Dfc} - T_{Din}\right)\left(y_{Dout} - y_{Din}\right) \tag{5.1.100}$$

Rearranging will give,

$$T_{Dout} = -\varphi N_g \left(y_{Dout} - y_{Din}\right) + N_f \omega \frac{P_{in}^2}{T_{in}^2} \left(y_{Dout} - y_{Din}\right) - \left[T_{Di}\left(1-\aleph\right) + \aleph T_{Dfc}\right]\left(y_{Dout} - y_{Din}\right) \tag{5.1.101}$$

Recalling definition of dimensionless temperature by equation 5.1.102

$$T_{Dout} = \frac{T_{out}}{T_{in}} \quad 5.1.102$$

By similarity, dimensionless formation temperature can be obtained.

$$T_{Din} = \frac{T_{in}}{T_{in}} \quad 5.1.103$$

$$T_{Dfc} = \frac{T_{fc}}{T_{in}} \quad 5.1.104$$

5.1.2.1.7 The Flow-Conduit Inner Temperature and OHTC Calculation

At this point, all parameters needed to compute the fluid's temperature at elevation *y* can be obtained by calculation and by direct input, except for the inside conduit surface temperature. The computation of this temperature is based on ideas proposed by Satter (1965) and adapted by Willhite (1967). The latter computed casing's inside temperature of a steam injection well as a function of geothermal temperature. His procedure estimates overall heat transfer coefficient in an iterative manner. It has been modified in this work and extended to also calculate the tubing inside temperature.

The heat transferred to/from the formation to the formation-cement interface casing can be calculated by rearranging equation 5.1.73 to give equation 5.1.105 below.

$$q_{fc} = \frac{2\pi K_f \left(T_F - T_{fc}\right)\Delta L}{f(t)} \quad 5.1.105$$

The heat transferred through the tubing, the casing and the cement sheet is expressed under assumption that heat flows from formation to fluid stream (although the inverse would occur), by equations 5.1.106, 5.1.107, and 5.1.108, respectively.

$$q_{tub} = \frac{2\pi k_{tub}\left(T_{to} - T_{ti}\right)}{\ln\left(r_{to}/r_{ti}\right)}\Delta L \quad 5.1.106$$

$$q_{cas} = \frac{2\pi k_{cas}\left(T_{co} - T_{ci}\right)}{\ln\left(r_{co}/r_{ci}\right)}\Delta L \qquad 5.1.107$$

$$q_{cem} = \frac{2\pi k_{cem}\left(T_{cemo} - T_{cemi}\right)}{\ln\left(r_{cemo}/r_{cemi}\right)}\Delta L \qquad 5.1.108$$

Assuming that the cement and the casing are tight enough for inside cement temperature to be equal to outside casing temperature and for the cement sheet inner radius to be equal to casing outer radius, equation 5.1.108 can be rewritten as equation 5.1.109.

$$q_{cem} = \frac{2\pi k_{cem}\left(T_{cemo} - T_{co}\right)}{\ln\left(r_{cemo}/r_{co}\right)}\Delta L \qquad 5.1.109$$

The heat transferred from the cement-formation interface to the fluid flowing in the tubing is calculated with equation5.1.110.

$$q = 2\pi r_{over} U_{over}\left(T_{fc} - T_f\right)\Delta L \qquad 5.1.110$$

Where, T_f is the fluid temperature in the well at the corresponding depth.

Because of the assumption of steady state heat transfer, the same amount of heat that flows through cement will flow through casing and tubing. This is equal to the amount of heat that flows from the formation to the fluid and vice–versa. Therefore equation 5.1.111 holds.

$$q_{cem} = q_{cas} = q_{tub} = q \qquad 5.1.111$$

The casing inside temperature can be calculated as a function of the overall heat transfer coefficient, by combining equations 5.1.110 , 5.1.108, 5.1.107 and 5.1.111.

$$T_{ci} = T_{fc} + \left(\frac{\ln\frac{r_{cemo}}{r_{co}}}{K_{cem}} + ... + \frac{\ln\frac{r_{co}}{r_{ci}}}{K_{cas}}\right) r_{over} U_{over}\left(T_f - T_{fc}\right) \qquad 5.1.112$$

Still because of the assumption of steady state conduction within wellbore components, equation 5.1.110 can be equated to equation 5.1.105 and be rearranged to obtain equation 5.1.113 in terms of the overall heat transfer coefficient.

$$T_{fc} = \frac{T_f f(t) + \frac{K_f}{r_{over} U_{over}} T_F}{f(t) + \frac{K_f}{r_{over} U_{over}}} \qquad 5.1.113$$

The calculation follows an iterative procedure in each segment, starting from wellhead to bottomhole. The iterations are conducted within two loops. This is clearly explained in the chapter 6, which elaborates on the implementation of the models into the simulation codes.

5.1.2.1.8 The Temperature in the Annulus

In light of the steady state heat transfer assumption through the wellbore components, the same amount of heat that flows from the tubing to the annulus will flow from the annulus to the casing and vice-versa. Therefore equation 5.1.114 holds.

$$(h_c + h_r) A_{at} \cdot (T_{ot} - T_{an}) = (h_c + h_r) A_{ac} \cdot (T_{an} - T_{ic}) \qquad 5.1.114$$

Developing equation 5.1.114 and rearranging it, gives equation 5.1.115. This equation suggests that the temperature in the annulus is a function of the average of the surface area of the casing inside and the surface area of the center of the annulus, and that of the average of the surface area of the tubing outside and the surface area of the center of the annulus. These average surface areas are calculated as log-average surface areas as earlier presented. It is assumed that the center of the annulus is the midway between tubing outer surface and casing inner surface.

$$T_{an} = \frac{A_{at} T_{ot} + A_{ac} T_{ic}}{A_{at} + A_{ac}} \qquad 5.1.115$$

5.1.2.2 Heat Transfer by Unsteady State Conduction through Surrounding Rock

The estimation of the temperature of the surrounding rock with time, assumes that the well behaves as an infinite cylinder source. Eppelbaum and Kutasov (2006) pointed out that for practical purposes, a cylinder whose length is five times or more its diameter could be treated as an infinite cylinder. They went on to proposed an expression for the determination of the effective wellbore radius to account for the effect of thermal resistance on the heater's wall temperature., By analogy, thermal resistance in this work is accounted for with the overall heat transfer coefficient. It informs on the propensity of the wellbore components between the flowing fluid and the rock (Tubing, annulus, casing, cement) to transfer heat.

5.1.2.2.1 Radius of Thermal Influence

In theory, a cylindrical source with constant heat flow rate affects the temperature field of formations at a long radial distance which in practice has a limit referred to as the radius of thermal influence (Eppelbaum and Kutasov, 2006) at which temperature $T\ (r_{in},\ t)$ approaches undisturbed geothermal temperature. In order to avoid uncertainty related to temperatures' estimates, Eppelbaum and Kutasov (2006) suggested that the radius of thermal influence should not depend on temperature difference between $T(r_{in},\ t)$, the temperature at radius r in the rock and T_i, the temperature in the wellbore. Therefore they proposed the thermal balance method to calculate the radius of thermal influence (r_{in}) by analogy to the material balance applied for calculation of drainage radius in pressure well testing. The dimensionless radius of thermal influence (R_{in}) can be obtained from equation 5.1.116 below, of which full derivation is presented in Appendix B.2.

$$t_D = \frac{1}{4}\frac{R_{in}^2 - 2\ln(R_{in}) - 1}{\ln(R_{in})}\ln\left[1 + \left(c - \frac{1}{a + \sqrt{t_D}}\right)\sqrt{t_D}\right] \qquad 5.1.116$$

Equation 5.1.116 is solved by iteration for R_{in} as a function of the value of t_D obtained from equation 5.1.117.

$$t_D = \alpha_R \frac{t}{r_w^2} = \frac{\lambda t}{C_{pR}\rho_R r_w^2} \qquad 5.1.117$$

Then the radius of influence r_{in} can be calculated with equation 5.1.118.

$$r_{in} = R_{in} r_w \qquad 5.1.118$$

5.1.2.2.2 Temperature of the Surrounding Rock

The temperature in the surrounding rock will change as injection proceeds. Such change would be marked at the near the wellbore and would extend through the region within radius of thermal influence in a less pronounced manner. This change would not be permanent as the rock and wellbore components would heat back to asymptotically approach undisturbed geothermal temperature some time after injection has ceased. Notwithstanding, it is important to have a picture of how temperature will evolve in the region within radius of thermal influence. This will affect fluid temperature in the wellbore and hence its pressure as well as its thermodynamic and transport properties. Eppelbaum and Kutasov (2006) proposed equation 5.1.119 to calculate wellbore wall temperature for an infinite cylindrical heater, which holds under consideration of the wellbore as an infinite cylinder having heat interaction with the formation, through well components.

$$T_w = T(t, r_w) = T_i + \frac{q}{2\pi\lambda} \Delta L \ln\left[1 + \left(c - \frac{1}{a + \sqrt{t_D}}\right)\sqrt{t_D}\right] \quad 5.1.119$$

In terms of rock-cement interface temperature T_{fc} equation 5.1.119 will yield 5.1.120.

$$T_{fc} = T(t, r_{fc}) = T_i + \frac{q}{2\pi\lambda} \Delta L \ln\left[1 + \left(c - \frac{1}{a + \sqrt{t_D}}\right)\sqrt{t_D}\right] \quad 5.1.120$$

Where,

q is the heat flow rate per unit length

T_i is the undisturbed rock temperature

They further proposed two expressions for dimensionless time at radius r within the formation (5.1.121 and 5.1.122). (5.1.123).

$$T_{Dr} = 1 - \frac{\ln r_D}{\ln R_{in}} \quad 5.1.121$$

$$T_{Dr} = \frac{T(r,t) - T_i}{T_{fc} - T_i} \quad 5.1.122$$

Rearranging equating equation 5.1.122 will yield 5.1.123, with $T_{Dr}(r_D, t_D)$ obtained with equation 5.1.121. Equation 5.1.123 is used for calculating rock temperature at radius *r* as a function of the fluid's temperature T_i and the temperature at the formation-cement interface T_{fc}.

$$T(r,t) = (T_{fc} - T_i)T_{Dr}(r_D, t_D) + T_i \qquad 5.1.123$$

Where,

$$r_D = \frac{r}{r_w} \qquad 5.1.124$$

Equation 5.1.121 and 5.1.123 suggest that once R_{in} , is known, $T(r,t)$ the temperature at radius *r* within the rock, can be calculated.

5.2 Isolated System Behavior

This system assumes no energy exchange between the gas and the surroundings implying absence of any external energy source and/or sink. The work is taken from and/or stored as the internal energy of the gas and all converted to kinetic energy and/or heat resulting in an adiabatic cooling/heating process. Two major cases can be considered: the non-isothermal and the isothermal cases respectively.

5.2.1 The Non-Isothermal Case

Adiabatic non-isothermal situation is approached at high injection rates, where the fluid flows so fast that the heat absorbed by it is not significant as compared with the available heat in the formation. The analysis of the fluid temperature and pressure under this condition is based on the first law of thermodynamics, stating conservation of energy and given by equation 5.2.1 below.

$$dU = Q - PdV \qquad 5.2.1$$

Where,

U is the internal energy and

Q is heat energy

Under assumption of adiabatic process, the heat energy equates to zero $(Q = 0)$. As a consequence, equation 5.2.1 reduces to 5.2.2.

$$dU = -PdV \tag{5.2.2}$$

The term on the right hand side of equation 5.2.2 can be expressed as a function of the heat capacity at constant volume (C_V), by equation 5.2.3.

$$-PdV = C_V dT \tag{5.2.3}$$

Recalling the ideal gas law in equation 5.2.4

$$PV = nRT \tag{5.2.4}$$

Putting equation 5.2.4 into 5.2.3 and rearranging gives 5.2.5.

$$-Pd\left(\frac{nRT}{P}\right) = C_v dT \leftrightarrow -nRPd\left(\frac{T}{P}\right) = C_v dT \tag{5.2.5}$$

Upon expansion, equation 5.2.5 yields 5.2.6

$$\frac{dT}{T} = \frac{dP}{P}\left(\frac{nR}{C_V + nR}\right) \tag{5.2.6}$$

The heat capacity at constant volume C_V and that at constant pressure C_P can be related through equations 5.2.7 and 5.2.8.

$$C_p - C_V = nR \tag{5.2.7}$$

$$\frac{C_p}{C_V} = \gamma \tag{5.2.8}$$

Putting 5.2.7 and 5.2.8 into 5.2.6 yields equation 5.2.9

$$\frac{dT}{T}=\frac{dp}{p}\frac{C_p-C_V}{C_p}=\frac{dP}{P}\left(1-\frac{1}{\gamma}\right)=\frac{dP}{P}\left(\frac{\gamma-1}{\gamma}\right) \qquad 5.2.9$$

Integrating and rearranging equation 5.2.9 gives equation 5.2.10 .

$$\frac{p^{\left(\frac{\gamma-1}{\gamma}\right)}}{T}=Constant \qquad 5.2.10$$

Equation 5.2.10 can be used to determine the bottom hole pressure knowing tubing head pressure and temperature and bottom hole temperature for an ideal gas. This could serve for quick check in low-pressure and high temperature operations applicable for injection in depleted gas or oil reservoirs.

In the case of real gases at high pressure and/or low temperature, the Joule-Thomson effect will contribute to heating/cooling of the fluid.

Let's recall that equation 5.2.11 will hold for an isolated system. This implies that there is no heat exchange between fluid and surrounding rock formation.

$$\delta q\approx 0 \qquad 5.2.11$$

Therefore equation 4.4.29 will reduce to 5.2.12., which could be used to estimate fluid temperature at the outlet of a segment in gas injection well under adiabatic condition, approached whenever the flow-conduit is isolated with insulating materials on its outer or inner surface.

$$dT_D=-\varphi N_g dy_D+\omega N_f\frac{P_{in}^2}{T_{in}^2}dy_D \qquad 5.2.12$$

5.2.2 The Isothermal Case

This is the case where the fluid temperature is considered to be constant along the well. It would be unrealistic for an injection well. However it can provide a good examination of the prediction of thermodynamic behavior of the fluid in the well. It could help to obtain more insight in the ability of the codes to predict the thermodynamic behavior of the fluid by comparing its predictions with those of the other models in which temperature changes along the well. The pressure calculation in this case can be performed with equation 5.2.13. This equation is a simplification of equation 4.3.16 to account for uniform along the well.

$$\frac{dP}{P} = -N_g \frac{dy}{L} + N_f \frac{P_{in}^2}{P^2} \frac{dy}{L} \qquad 5.2.13$$

5.3 Heat Loss or Heat Taken Up Due to Phase Change

The theoretical approach adopted in this work views enthalpy of vaporization as the heat extracted from the surroundings to increase the fluid's entropy when the liquid is vaporized. Conversely, the enthalpy of condensation will be taken as the heat released to the surroundings as a consequence of vapor condensing. For equilibrium to persist between liquid and vapor, which is the case at different boiling temperature T_{bt}, the change in chemical potential and hence the change in Gibbs free energy should equate to zero (Atkins, 1986, P.139) as in equation 5.3.1.

$$\Delta G_{T_{bt}} = \Delta H_{T_{bt}} - \Delta S_{T_{bt}} = 0 \qquad 5.3.1$$

Where,

$G$is the Gibbs free energy, H is the enthalpy, S the entropy and T_{bt} the boiling points.

Rearranging equation 5.3.1 will yield equation 5.3.2 in terms of enthalpy and entropy.

The change of entropy upon vaporization or condensation can be defined by equation 5.3.2 (Atkins, 1986, P.141) which upon rearranging will result in equation 5.3.3.

$$\Delta S = \frac{\Delta H_{pc}}{T_{bt}} \qquad 5.3.2$$

$$\Delta H_{pc} = \left(S_{vap} - S_{Liq}\right) T_{bt} \qquad 5.3.3$$

With entropy of phase change ΔS_{pc}, defined by equation 5.3.4.

$$\Delta S_{pc} = S_v - S_L \qquad 5.3.4$$

Where,

The indices *V* and *L* and *pc* stand for vapor, liquid and phase-change respectively.

This temperature change will result to heating or cooling. The heat will be transferred by forced convection, inducing temperature change in the flow conduit, from which the heat will propagate radially through the wellbore components into the formation.

5.3.1 Phase Identification

The identification of the flowing phase below critical point (temperature below critical temperature and pressure below critical pressure), is based on the comparison of the vapour pressure obtained from the saturation pressures table with the prevailing pressure in each segment of the well. Fluid at pressures higher than vapor pressure is in liquid phase and fluid at pressures smaller than vapor pressure is in the gaseous phase.

Above critical point, where temperature is higher than critical temperature and pressure higher than critical pressure, the fluid is identified as supercritical fluid.

The other scenarios involve either only pressure higher than critical pressure or only temperature higher than critical temperature. In this situation, above critical temperature, when pressure is lower than critical pressure the fluid is also identified as gas whereas above critical pressure, when temperature is lower than critical temperature, fluid will always be liquid.

6. THE SIMULATION CODES AND THE GRAPHICAL USER INTERFACE

As earlier introduced, simulation codes have been written in Microsoft Visual Studio with Visual Basic.Net language to implement the solutions of the models. A user-friendly graphical interface has also been developed to encapsulate the solution codes and permit easy use and deployment of the codes. This chapter elaborates on how the most important aspects of the work have been implemented . It presents the workflows implemented in the computation codes and the development of the graphical user interface.

6.1 Thermodynamic and Transport Properties Tables

The thermodynamic and transport properties tables are read from Microsoft EXCEL spreadsheets to create XML data files with a program written for the purpose. The simulation codes include a module to read the data from the XML data files during runtime. Virtual tables are created. Thermodynamic and transport properties are read from the tables at prevailing temperature and pressure in each well segment. The sequence for building the runtime virtual tables is summarized in Figure 6.1 below.

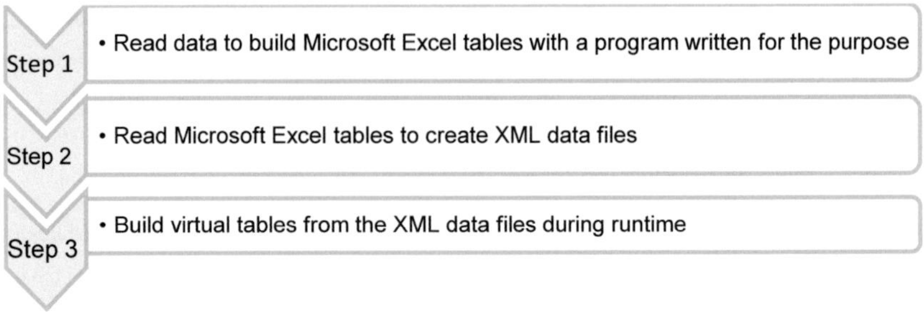

Figure 6.1*: Sequence for building virtual thermodynamic and transport properties tables*

6.2 Pressure and Temperature Calculation

Pressure calculation is performed only on the flowing fluid because the wellbore is a closed system. In the case of temperature, it is important to account for the thermal interactions between the flowing fluid and the surrounding rock. Surrounding rock would behave as a heat source (when its temperature is higher than fluid's temperature) or as a heat sink (when its temperature is lower than that of the fluid.). As such the temperature of the wellbore components (tubing, casing, annulus and cement) and that of the formation are also calculated to account for changes with time.

6.2.1 Flowing Fluid Temperature and Pressure

The calculation of pressure and temperature follow a marching algorithm represented by Figure 6.2 below. The well is divided into segments of length determined by a stringent validation criterion. This criterion imposes that the relative difference between the Z-factor at the inlet of the segment and that at its outlet should not exceed 0.1%. The fluid properties (density, enthalpy, viscosity, thermal conductivity, etc.) and the pressure and temperature along the segment are first assumed to be constant and equal to those at the inlet. These are used to calculate the temperature at the outlet of the segment (which is considered to be the temperature of the next segment). Then the new temperature and inlet pressure are used to read thermodynamic and transport properties from the tables. Thereafter the segment outlet pressure is calculated with this temperature and the newly read properties. These steps are repeated from one segment to the other going from the wellhead to the bottom of the hole.

The temperature calculation considers four cases: the isothermal case; the case of linear temperature profile corresponding to the geothermal temperature profile; and the case of rigorous temperature calculation based on energy balance in each well segment. There are two instances of the latter case:

the adiabatic and the non-adiabatic instances, which for ease of identification are also considered as two distinct cases. In the adiabatic instance, the wellbore is assumed to be thermally insulated from the surrounding rock. Therefore there is no heat exchange between the fluid and the rock. The non-adiabatic instance accounts for the heat exchange between the fluid and the surrounding rock in the energy balance equation.

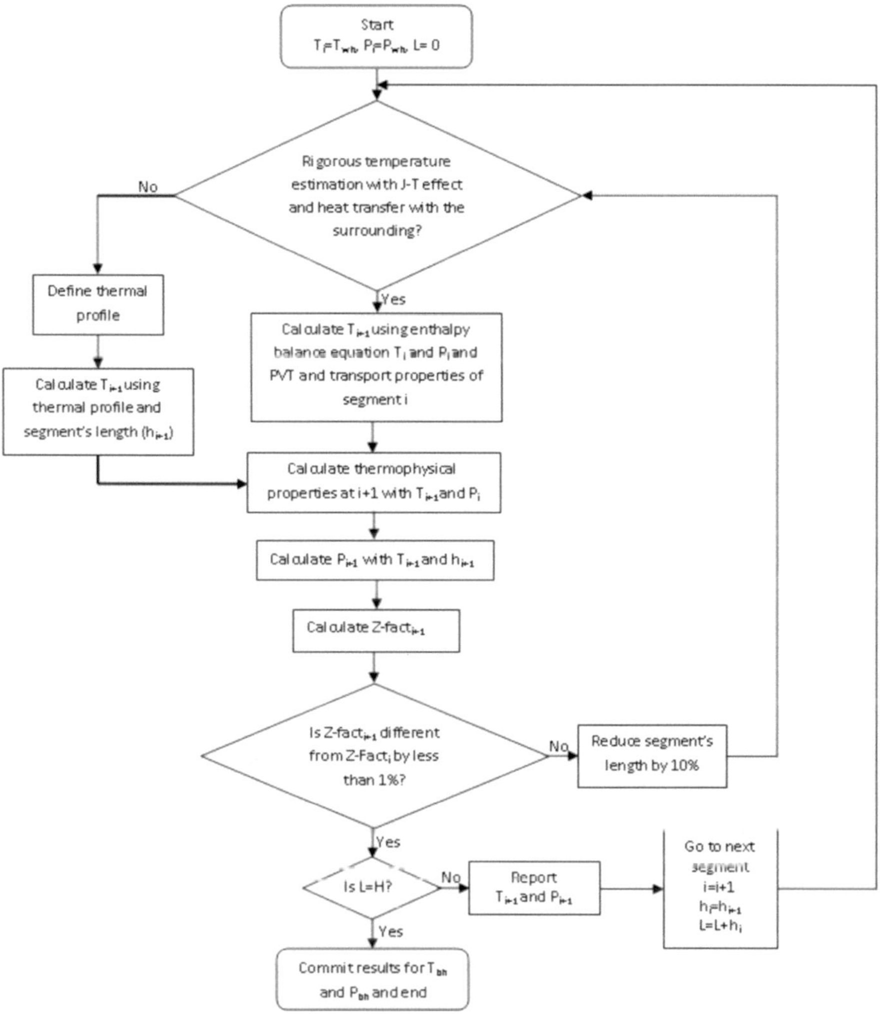

Figure 6.2: *Marching Algorithm for calculating temperature and pressure along the well*

6.2.2 Tubing, Casing, Rock-Cement Interface and Annulus Temperature

The flow chart consists of two loops. The first loop is embedded into the second one and it calculates the inner casing temperature. The second loop which also constitutes the main loop calculates the inner tubing temperature. Once inner casing and inner tubing temperatures are known the outer tubing and the annular temperatures are calculated.

At the inlet of a well segment, iteration is initiated by assuming the inner tubing temperature to be the average of fluid inlet temperature and geothermal temperature whereas the inner casing temperature is considered to be equal to geothermal temperature at corresponding depth. Then, in the first loop (for inner casing temperature), the overall heat transfer coefficient is computed using equation 5.1.65. After that, the inner casing temperature is computed as a function of the assumed inner tubing temperature, using equations 5.1.113 and 5.1.113. The iteration continues until the relative difference between assumed and computed inner casing temperatures is smaller than 0.1%. Thereafter, the fluid temperature at the outlet of the segment and the amount of heat taken up or loosed by the fluid are calculated as a function of the overall heat transfer coefficient and the film transfer coefficient (using equation 5.1.58) respectively. These are subsequently used to calculate the inner tubing temperature. The next inner tubing temperature is set to the average of the newly calculated value and the previous value. The loop iterates until the relative difference between current and previous values of inner tubing temperature is also smaller than or equal to 0.1%. Figure 6.3 shows the schematic representation of the algorithm to calculate OHTC, casing inside temperature and tubing inside temperature. It also calculates tubing outer temperature and the temperature in the annulus.

6.2.3 Surrounding Rock Temperature Calculation

The formation temperature is calculated based on equation 5.1.123. It based on an iterative algorithm that is closely dependent on the rock-cement interface temperature calculation. It iterates in a loop for the temperature at radius r and time t, until the relative difference between the radial distance at which the temperature is calculated and the radius of thermal influence is less than 0.1%.The steps can be summarized as follows:

1. Calculate the rock-cement interface temperature and the radius of thermal influence
2. Compute the temperature in the rock at radial distances increasing by 10% of the difference between the radius of thermal influence and the last radial distance at which the temperature has been calculated.

3. Loop until the relative difference between the radial distance at which the temperature is calculated and the radius of thermal influence is less than 0.1%.

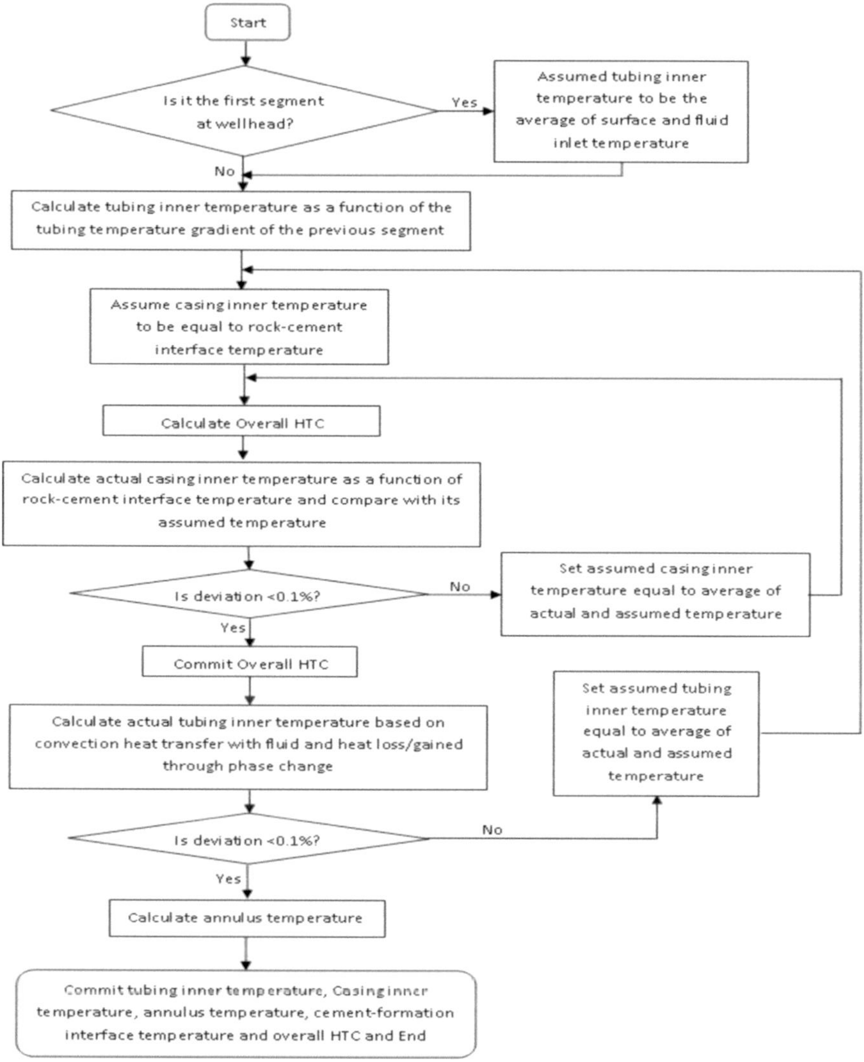

***Figure 6.3**: Flow chart to calculate overall heat transfer coefficient, inner tubing temperature, inner casing temperature rock-cement interface and annulus temperature*

6.3 Phase Change Check

The flow chart to check if the gas phase will change in a segment of the well is presented in Figure 6.4. It takes into account condensation/evaporation processes and their incidence on the fluid temperature in the well segment. It is executed only when temperatures (gas and rock) are below gas critical temperature or when there are possibilities that gas temperature falls below its critical temperature due to cooling by the surrounding rock. It can be summarized as follows:

1. Assume segment's outlet temperature to be equal to geothermal temperature in case the non-adiabatic and adiabatic processes are considered. For the isothermal process, the fluid temperature in the segment will still be the temperature at the inlet and for the geothermal case the fluid temperature will be equal to the geothermal temperature.
2. Calculate the segment's outlet pressure based assumed outlet temperature.
3. Calculate the vapor pressure corresponding to the inlet temperature in the segment.
4. Calculate vapour pressure corresponding to assumed outlet temperature.
5. Subtract inlet and outlet vapour pressure from inlet and outlet pressure respectively.
6. Compare the sign of the results of the subtractions. If the signs are different then there is possibility for phase change.
7. In there is possibility for phase change, reduce the segment's length iteratively until there is no possibility for phase change. If the length gets below 1.0 meter, consider that the gas phase will definitely change in the segment.
8. If there will definitely be phase change, set back segment's length equal to initial length before the start of iterations and set outlet temperature equal to inlet temperature to indicate that temperature will not also change in this segment. If at a particular segment's length it is found out that there is no possibility of phase change, set the segment's length equal to this length and exit the loop and continue with the loops for temperature calculation to obtain the segment's outlet temperature.
9. In the case of phase change, for evaporation, energy is taken from the surrounding. The heat extracted from the formation propagates through well components up to the gas and results in temperature drop in the rock. When there is rather condensation, heat is released to the surrounding. It propagates through the well components up to the formation and heats it to increase the temperature of the surrounding rock. In all the cases the temperature change in the rock is not significant.

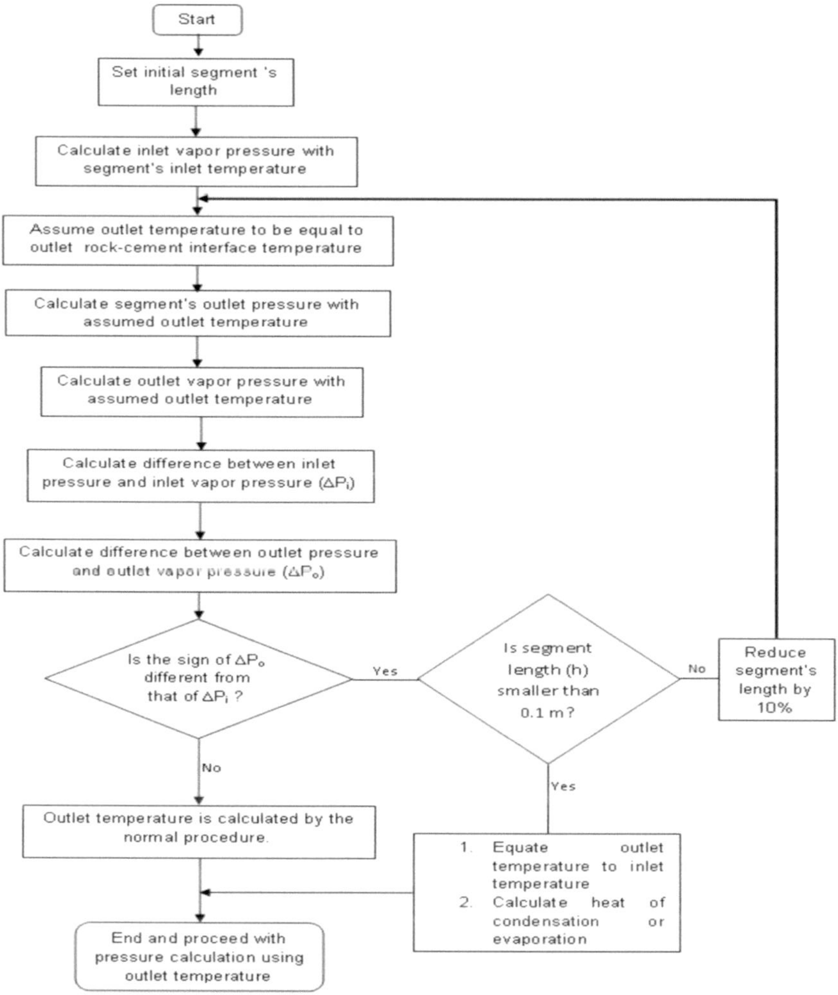

***Figure 6.4**: Flow chart to check if the gas phase will change in a segment of the well and calculate temperature accordingly*

6.4 Phase Identification

In the simulation codes, the identification of the flowing phase below critical point (temperature below critical temperature and pressure below critical pressure), is based on the comparison of the vapour pressure read from the saturation pressures' table, with the prevailing pressure in each segment of the well. Fluid at pressures higher than vapour pressure is in liquid phase and fluid at pressures smaller than vapor pressure is in the gaseous phase.

Above critical point, where temperature is higher than critical temperature and pressure higher than critical pressure, the fluid is identified as supercritical fluid.

The other scenarios involve either only pressure higher than critical pressure or only temperature higher than critical temperature. In this situation, above critical temperature, when pressure is lower than critical pressure the fluid is also identified as gas whereas above critical pressure, when temperature is lower than critical temperature, fluid will always be liquid.

Figure 6.5 gives the schematic representation of the flow chart used for phase identification. It is based on Figure 6.6 which shows the phase diagram of CO_2.

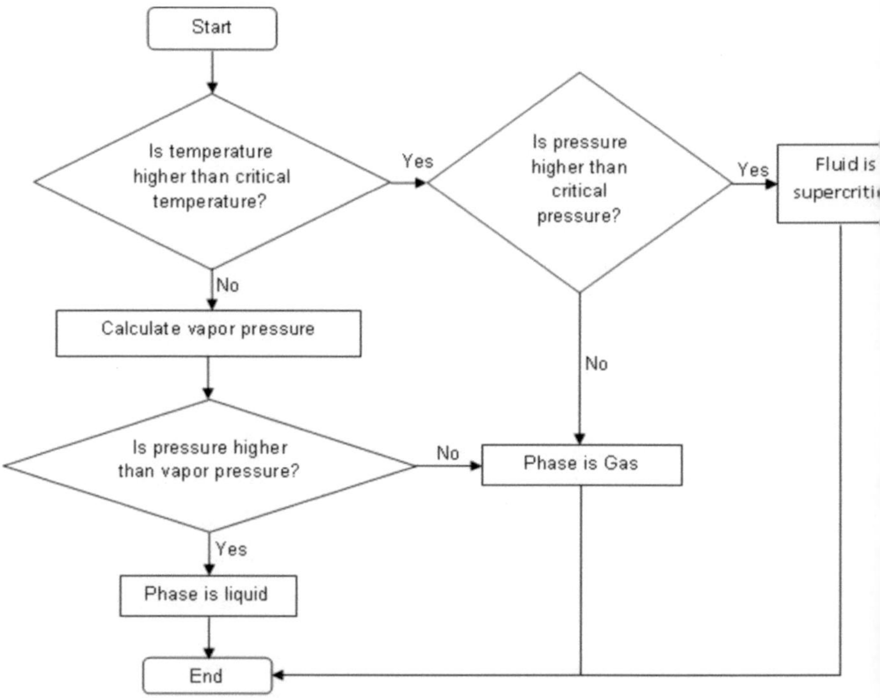

***Figure 6.5**: Flow chart to identify the phase of the gas in a segment of the well*

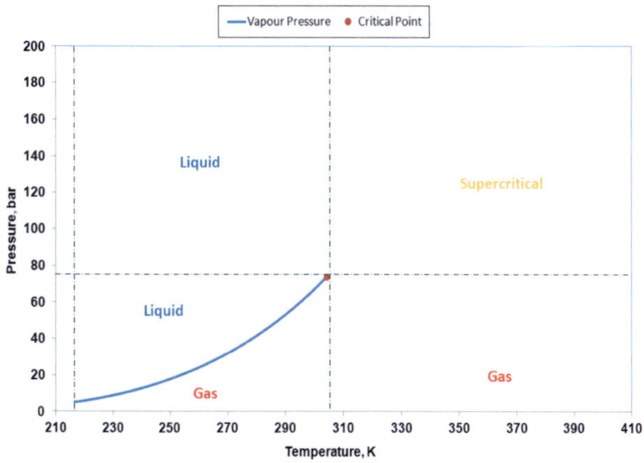

***Figure 6.6**: CO_2 Phase diagram used to guide phase identification during simulations*

6.5 The Graphical User Interface (GUI)

As earlier introduced, a user-friendly GUI has been developed to ease deployment and use of the codes. The interface is made up of five windows corresponding to each sequence of data input and a sixth window for simulation launching and data output. These windows are contained in a MDI parent container window. Figure 6.7 below shows the sequences for data input, simulation and data output with their corresponding simulator windows.

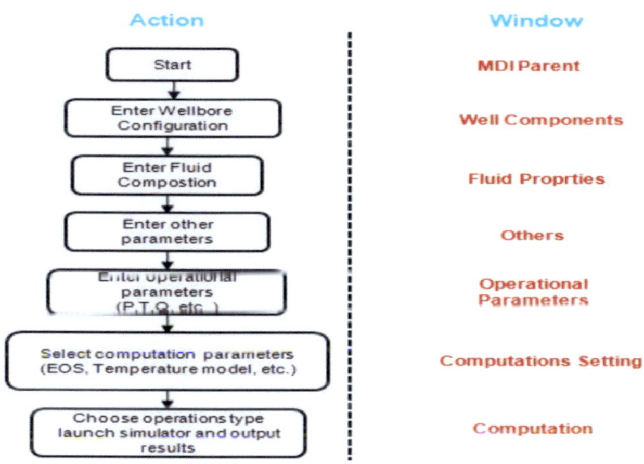

***Figure 6.7**: Sequence for data input, simulation and data output with their corresponding windows*

The data is output in text files with .txt format. It can readily be exported in Microsoft Excel for processing. Figure 6.8 to Figure 6.14 present the different windows with their data and information requirements. All input data and results are in SI units.

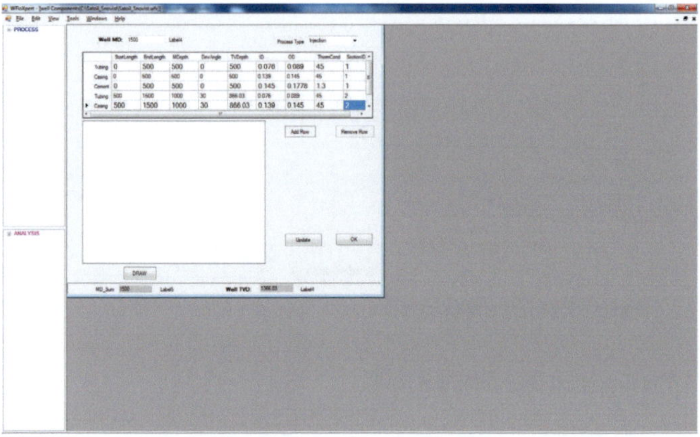

***Figure 6.8**: MDI parent window containing the wellbore Component's window*

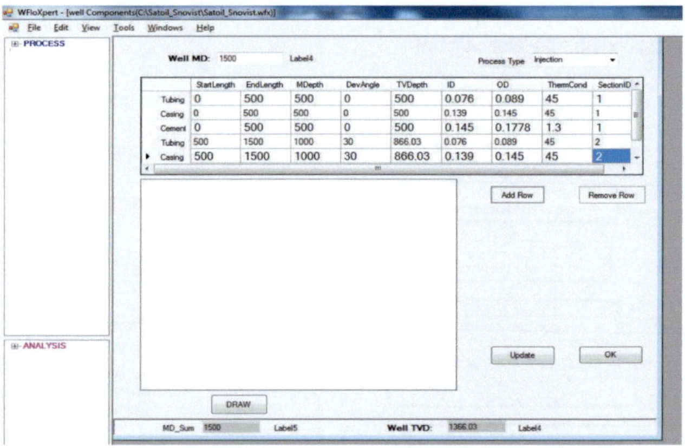

***Figure 6.9**: An instance of the wellbore components window showing tubing, casing and cement input data for a well made up of 2 sections with a 30° deviation angle in the second section*

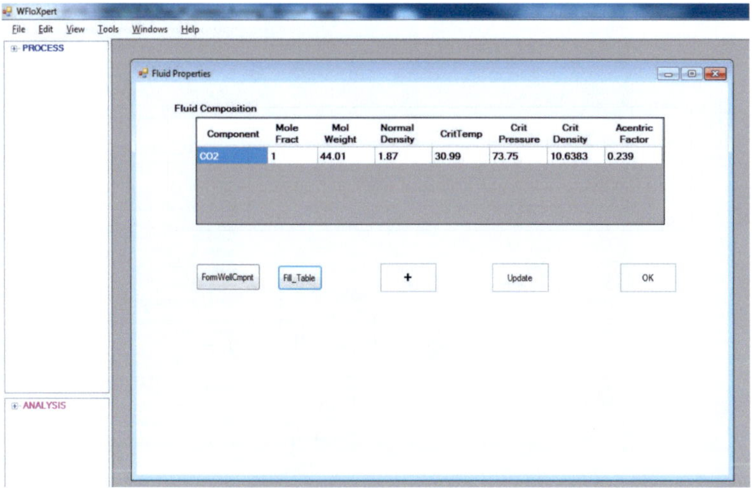

***Figure 6.10**: An instance of the fluid properties' window with entries of CO_2 properties*

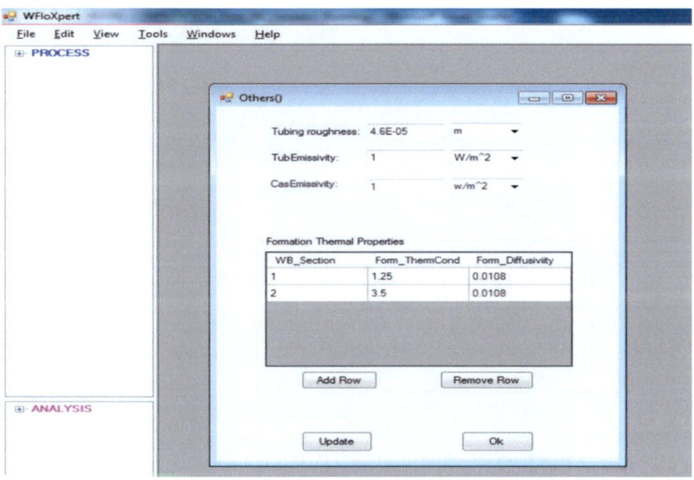

***Figure 6.11**: An instance of the window for entering other (miscellaneous) parameters showing thermal conductivities of 1.25 and 3.5 W/mK for well section 1 and 2 respectively*

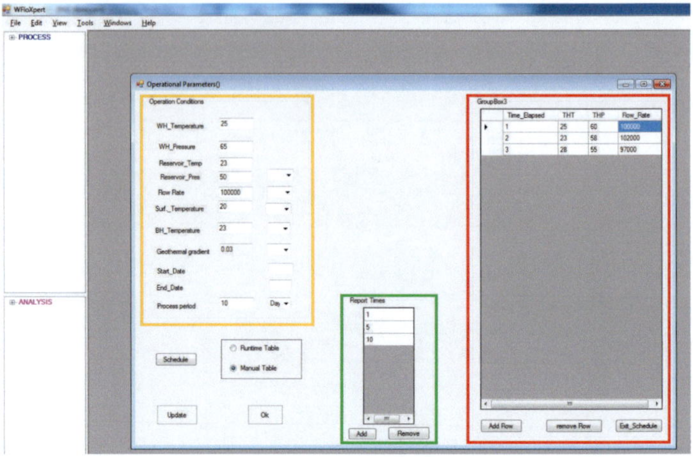

***Figure 6.12**: An instance the operational parameters windows showing the different way of scheduling the operations (for same FR, THP and THT, in the yellow frame, for scheduling with different FR, THT and THP in the red frame) and days for reports printing (in the green frame)*

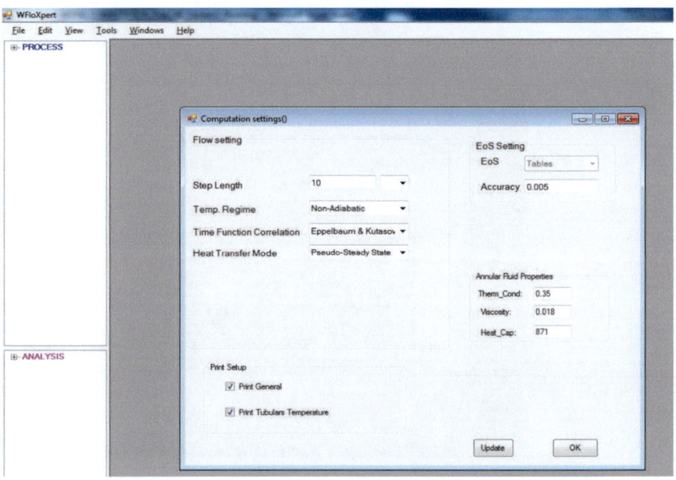

***Figure 6.13**: An instance of the settings window where the computation settings are selected (e.g. EOS, initial length of the well segments and parameters for heat transmission in the surrounding rock)*

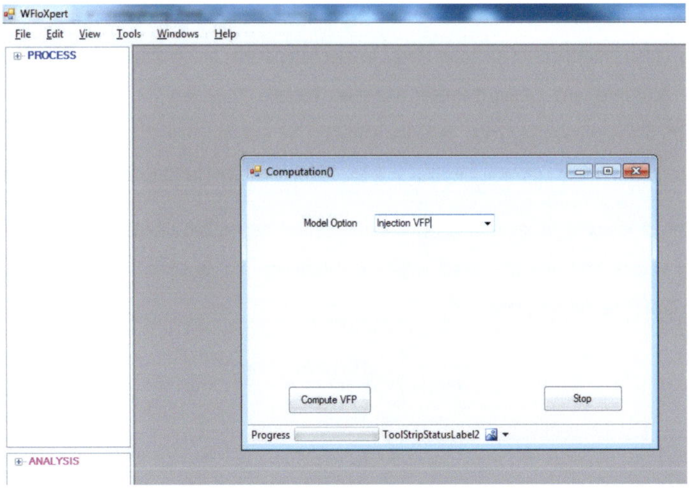

***Figure 6.14**: An instance of the computation window in which the process to simulate is selected (e.g. injection VFP) the simulation are launched and the results are output*

7. VALIDATION OF THE MODELS AND THE PROCEDURE

The validation of the model and procedure developed in this work would require comparison of the simulation results with field measured data or data from laboratory experiments. Field measured data from a recent demonstration project on CO_2 injection in a saline aquifer in Germany (Ketzin) and a commercial scale CO_2 injection pilot project in a saline aquifer in Norway (Snoehvit) is available. This field data is compared with the simulation results.

The measured data in the injection wells consists of one point measurement of pressure and temperature. Although the measured points are at different depths for the two wells (Ketzin & Snoehvit), more extensive test on the model's capabilities along the wells might be obtained with several measurement points. To this end, a further comparison of the simulation results with measured field data in a production well is also performed. This data have been used by Cronshaw and Boiling (1982) in validating their numerical model for non-isothermal flow of carbon dioxide in wellbores.

The agreement of procedure and models with theory is examined by comparing the performance of the temperature models and by highlighting the similarities of the main model (non-adiabatic model) with the others. Further examination of the models' agreement with theory is performed by analyzing the sensitivity of the predictions of the main model to injection parameters (wellhead temperature, wellhead pressure, flow rate, and tubing diameter and inner surface roughness), injection time and thermal conductivity of the surrounding rock. In all the simulations it was assumed that the increase of reservoir pressure with time will not prevent injection at selected rates and wellhead temperature and pressure.

The values of temperature and pressure at measurement depths are tabulated and the deviation of simulated values from the measured values is calculated and analyzed. The percentage of the deviation is obtained with equation 7.1.1

$$dev(\%) = \frac{Abs(Obs - Sim)}{Obs} * 100 \qquad 7.1.1$$

Where,

Obs is the measured value, Sim is the simulated value and dev is the percentage deviation

7.1 Comparison with Measured data of the Injection Wells

7.1.1 The Ketzin Well

Henninges et al. (2010) presented the Ketzin project and the injection well. The reservoir is in a sandstone layer of the Triassic Stuttgart Formation. It is located at approximately 630 meters depth. The initial temperature is around 33.5 °C and the initial pressure around 61 Bar. The injection well was drilled to the depth of about 800 meters. The temperature and pressure measurements have been taken at the measured and true vertical depth of 549 meters approximately. The wellbore configuration is presented in
Figure 7.1 below.

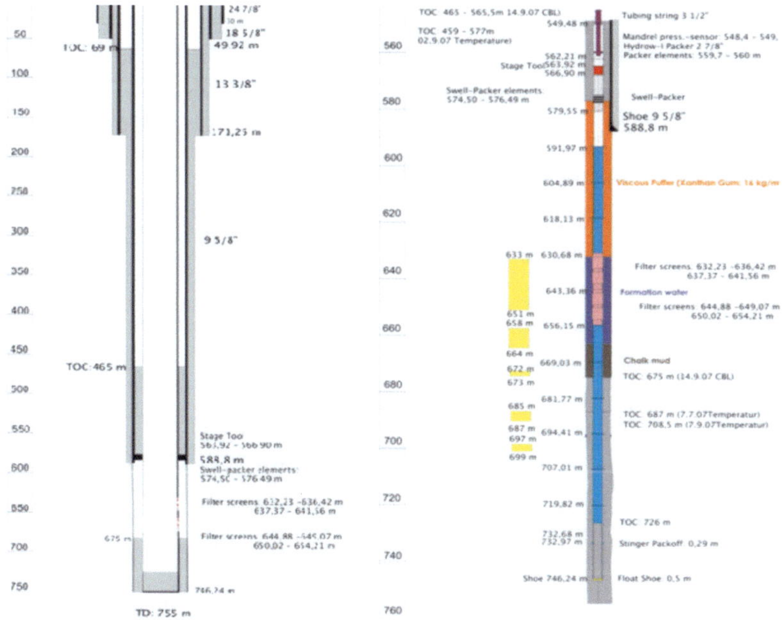

Figure 7.1*: Ketzin casing configuration in the wellbore (left); and Tubular in the reservoir portion of the well (right)*

7.1.1.1 Simulation Input Data

The longest tubing section has an outer diameter of 0.089 meter (3.5 inches) and inner diameter of 0.076 meters. This diameter is considered in the simulations. The tubing string stops at depth 562 meters. The wellbore configuration has been simplified in the simulations. Three sections are considered to account for different thermal conductivities as a result of the number of layers and different flow diameter accounting for the fact that the tubing stops at 562 meters and as such the flow in the remaining section of the wellbore is through the liner of 0.139 (5.5 inches) meters outer diameter and 0.121 meters inner diameter. The difference in thermal conductivities of the section has been considered by assigning a lower thermal conductivity to the cement layer of the uppermost section (because it has more casing layers) and highest thermal conductivity to that of the lowermost section (because it has less casing layers). The resulting configuration is presented in table 7.1 below.

The rock thermal conductivity and diffusivity have been assumed to have the highest uncertainty of all input parameters, and have, as such been used as the matching parameter. In sections 1 and 2 of the wellbore rock thermal conductivities of 3.0W/mK and 2.5W/mK and thermal diffusivities of 0.003m²/h and 0.0025m²/h respectively provided closest match of simulated values to measured ones. Whereas in section 3 a thermal conductivity value of 1.75 W/mK with thermal diffusivity of 0.0018 m²/h yielded reasonable extrapolation of temperature and pressure to the bottom of the well. These values of thermal conductivities are within the range reported by Norden et al. (2012) for a reservoir in Altensalzwedel area which is in the North German basin like Ketzin .

***Table 7.1**: Wellbore configuration and thermal conductivity data used for the Ketzin well*

Sections	End MD	Dev. Angle	ID	OD	TC	Item
	m	degrees	m	m	W/mK	
1	171	0	0.076	0.09	45.00	Tubing
			0.226	0.25	45.00	Casing
			0.245	0.31	1.30	Cement
2	562	0	0.076	0.09	45.00	Tubing
			0.226	0.25	45.00	Casing
			0.245	0.31	1.40	Cement
3	755	0	0.121	0.13	45.00	Tubing
			0.131	0.14	45.00	Casing
			0.139	0.22	1.50	Cement

Surface temperature was taken as 10°C based on information from DTS[11] measurements in the annulus. Geothermal gradient is taken as 0.037°C/m to give a temperature of 33.5°C reported as initial reservoir temperature at 630 m and a temperature of 37.9°C at the bottom of the wellbore at 755 m. The annulus thermal conductivity is an uncertain parameter. It was taken as 0.25 W/mK. Table 7.2 summarizes the wellhead temperature, pressure and flow-rate during the injection period.

***Table 7.2**: Input wellhead injection data for the Ketzin well*

Day	Temperature	Pressure	Flow Rate
	°C	Bar	m³/day
1	37.09	55.37	26375.00
2	30.27	57.22	29539.00
3	30.38	57.86	29220.00
4	31.90	58.36	26676.87
5	28.93	58.71	29746.40
6	29.84	58.63	26773.74
7	30.87	58.25	28149.16
8	26.71	58.85	36841

[11] Distributed Temperature Sensing

7.1.1.2 Results and discussion

Injection started on 30th June 2008. In the beginning there were no measurements in the wellbore. Temperature and pressure measurements started ten days later. Thereafter there were almost three days without injection. Because low volume of CO_2 was cumulatively injected with periods of injection and shut-in, it is assumed that the surrounding rock has recovered its temperature during the three consecutive shut-in days. Therefore the model was set up without accounting for eventual change in the temperature distribution in the surrounding rock caused by earlier injection. Eight straight injection days following this shut-in period were selected to test the models (see table 7.2). The wellhead pressure and temperature and flow rates were very close. This made sure that the Joule-Thomson effect would not be too different in the well and amongst the injection days. The data was measured on an hourly basis. Therefore daily average pressure, and flow rates were considered for the 24 hours of each injection day. However the temperature reported in the last injection hour was considered as daily measured temperature. This is justified by the fact that the change in the surrounding temperature and consequently the effect of the heat transfer between the flowing fluid and the surrounding rock are also function of the cumulative volume that flew in the well for a certain period.

Simulated and measured (at 549 meters measured and true vertical depth) values of pressure and temperature in the wellbore are compared in table 7.3. Further comparisons are presented in Figure 7.2 and Figure 7.3 for pressure and temperature respectively.

The percentage of the deviation of simulated values from the measured ones has been calculated with equation 7.1.1. The highest percentages are 1.34% for pressure and 3.32% for temperature. The high accuracy in pressure calculation is justified firstly by the stringent validation criteria in the well segments that force the errors to be minimized. Secondly the high accuracy of the models in calculating not only the thermophysical properties (density, viscosity, etc.) of the fluid in the well but also in calculating the temperature of the fluid further contributes to minimize the errors. The uncertainty on thermal properties of the surrounding rock and somehow on wellbore cement and tubular used to calculate temperature is higher than the uncertainty on density and viscosity which are used in calculating the pressure. Consequently, although the deviations of simulated temperature values from measured ones are small they are still higher than deviations of simulated pressure values from measured ones.

Table 7.3: *Comparison of measured and simulated pressure and temperature in Ketzin injection well*

Day	Pressure Bar			Temperature °C		
	Measured**	**Simulated**	**Deviation (%)**	**Measured****	**Simulated**	**Deviation (%)**
1	66.60	65.80	1.20	30.91	30.10	2.62
2	69.14	69.24	0.14	31.67	31.50	0.54
3	70.17	70.25	**0.11**	31.98	31.84	0.44
4	70.50	70.9	0.57	32.08	32.40	1.00
5	70.99	71.72	1.03	32.20	32.60	1.24
6	71.11	71.22	0.15	31.94	32.68	2.32
7	70.65	69.70	**1.34**	32.21	33.28	**3.32**
8	71.60	71.11	0.68	32.43	33.43	3.08

*** Measurement taken at 549 m measured and true vertical depths*

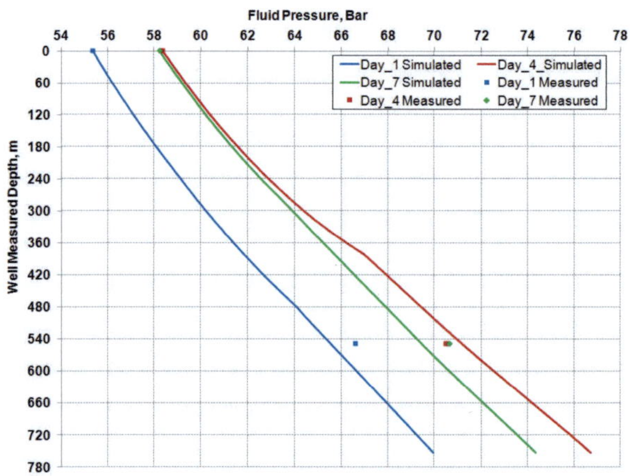

Figure 7.2: *Comparison of simulated pressures to measured pressure in the Ketzin well*

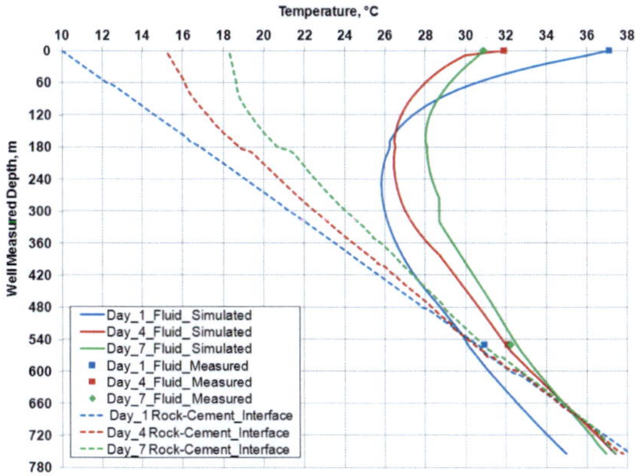

Figure 7.3: *Comparison of simulated temperatures to measured temperature and prediction of the rock-cement interface temperature in the Ketzin injection well*

Figure 7.3 shows the evolution of the rock-cement interface temperature with time. Initially (on day 1) the rock-cement interface temperature is assumed to be the undisturbed formation temperature and is proportional to the geothermal gradient. As injection goes on, because the gas in the wellbore is hotter than the surrounding rock, heat is transferred from the well to the rock. As a result, the temperatures of surrounding rock and that at the rock-cement interface and the wellbore components increase while that of the gas decreases.

An example of how the model predicts the propagation of the temperature in the surrounding rock is shown in Figure 7.4. This figure also shows how the radius of thermal influence (RTI) at 375 m depth is predicted to extend with time. Similar figures can be found in Appendix C (figure C1 and figure C2) for depths of 5 meters and 750 meters representing the top and the bottom of the well respectively.

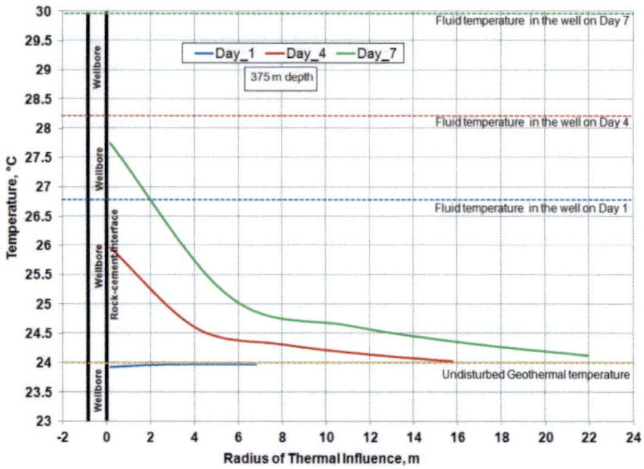

Figure 7.4: *Predictions of the evolution of rock temperature in the RTI at 375 m depth in Ketzin well*

The radius of thermal influence increases with time as injection goes on. Because the gas is hotter than the surrounding rock, it heats the rock. As a consequence, heat is transferred from the rock-cement interface radially inside the formation until the points where the temperature becomes asymptotic to the undisturbed geothermal temperature. The extent of the radius of thermal influence with time is proportional to the thermal conductivity of the rock and the difference between the fluid temperature and the undisturbed geothermal temperature. This point is extensively discussed in the section dedicated to sensitivity analyses. It is important to consider that there is rather a fast propagation of temperature in this case caused by the relatively high thermal conductivity considered as input (3.0 W/mK). For lower thermal conductivities slower propagation should be expected.

At this point it would be important to confirm the capabilities of the models in this work on different injection conditions and in a longer well that could also represent an oil or gas well. This is done with the Snøhvit well as earlier introduced and constitutes the topic of the next sections.

7.1.2 The Snøhvit Well

Snøhvit well is located offshore. The wellhead is situated at approximately 315 m TVD[12] below the sea level. The CO_2 is produced in the Melkøya field and transported through a pipeline of length approx. 153 Km. Injected CO_2 contains 1-5% hydrocarbon impurities, which are neglected in this study. CO_2

[12] True Verttical Depth

reaches wellhead at approximately 4°C, which would correspond to the average temperature of the sea water at this point. The reservoir is located at around 2333 meters TVD from the wellhead. Its initial average temperature and pressure are reported to be around 98°C and 216 bar (Statoil, 2009).

The wellbore configuration is presented in Figure7.5. The well has a length of approximately 2600 meters from the seafloor. The first section (944m) is deviated to a maximum of 8° and the remaining section of the well has a maximum deviation of 20°C. These result in a Through Vertical Depth of 2480 meters approximately.

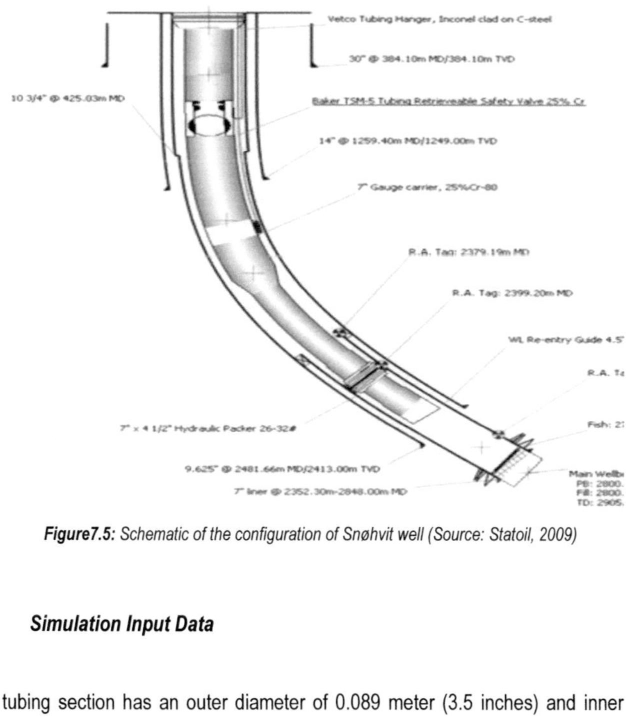

Figure7.5: *Schematic of the configuration of Snøhvit well (Source: Statoil, 2009)*

7.1.2.1 Simulation Input Data

The longest tubing section has an outer diameter of 0.089 meter (3.5 inches) and inner diameter of 0.076 meters. This diameter is considered in the simulations. The tubing string stops at depth 2165 meters. The wellbore configuration has been simplified in the simulations. Three sections have been considered to account for different thermal conductivities as a result of the different number of components layers (casing and cement) and different flow diameter accounting for the fact that the tubing stops at 2165 meters. As such the flow in the remaining section of the wellbore is through the liner of 0.168 meters (6.625 inches) outer diameter and 0.15 meters inner diameter. The difference in thermal conductivities of the sections has been recognized by assigning a lowest thermal conductivity to the cement layer of the uppermost section (because it has more casing layers) and highest thermal conductivity to that of the lowermost section (because it has less casing layers). The resulting configuration is presented in table 7.4 below.

As in the Ketzin's case rock thermal conductivity and diffusivity have been considered parameters with major uncertainties in the simulations. During simulations acceptable matches were achieved with rock thermal conductivity values between 1.0 and 1.5 W/mK and thermal diffusivity 0.001 m^2/hr assigned to all the three sections of the wellbore. The best match was achieved with the value of 1.25 W/mK. Unlike in the case of Ketzin, the three sections of the wellbore were assigned the same thermal conductivity value. This is reasonable because the well is located below the sea. It is very likely that most the rock are porous media with lowly to fairly consolidated sand. This justifies the values of rock thermal conductivities and thermal diffusivity that provided good temperature and pressure matches (Robertson, 1988).

***Table 7.4**: Wellbore configuration and thermal conductivity data used for Snøhvit well*

Sections	End MD	Dev. Angle	ID	OD	TC	Item
	m	degrees	m	m	W/mK	
1	944	8	0.076	0.09	45.00	Tubing
			0.220	0.24	45.00	Casing
			0.240	0.26	1.0	Cement
2	2165	20	0.076	0.09	45.00	Tubing
			0.220	0.24	45.00	Casing
			0.240	0.26	1.20	Cement
3	2590	20	0.150	0.17	45.00	Tubing
			0.171	0.18	45.00	Casing
			0.18	0.25	1.40	Cement

Moreover to achieve an early temperature stabilization in the well as indicated by measured data it was assumed that moving ground water was influencing temperature distribution in the section 1 and 2 of the wellbore. As such heat transported from depth prevented temperature from dropping very fast in section 1. Surrounding temperature at the wellhead is taken equal to wellhead gas injection temperature (4°C) corresponding to average sea temperature at that depth. Geothermal temperature is taken as 0.04°C/m to give 98°C reported as initial reservoir temperature at 2333 m TVD. The annulus thermal conductivity is an uncertain parameter. It was taken as 0.5 W/mK which is higher than in the

Ketzin case where it is assumed to be 0.25 W/mK. This is because Snøhvit is deeper and therefore annular fluid will be of much higher pressure and temperature compared with Ketzin, and therefore of higher thermal conductivity. Table 7.5 summarizes the wellhead temperature pressure and flow-rate during the injection period.

***Table 7.5**: Input wellhead injection data for Snøhvit well*

Days	Temperature	Pressure	Flow Rate
	°C	Bar	m³/day
1	4	105.00	447583.39
2	4	97.60	498215.11
3	4	96.34	502985.84
4	4	98.86	490512.53
5	4	98.24	498516.32
6	4	97.90	499351.45
7	4	98.92	496581.14
8	4	99.35	502558.47
9	4	101.02	435637.33
10	4	101.73	493062.85
11	4	100.11	520605.17

7.1.2.2 Results and discussion

Injection started on 18th April 2008. In the beginning there was a continuous alternation of injection and shut-in periods. There was also an obvious effect of the wellbore storage on pressure measurements in the wellbore. The wellbore storage effect disappeared on around the fourth day of injection and recorded values of pressure could be significantly higher than reservoir pressure. Thereafter there was a shut-in period of almost a day. Because low volumes of CO_2 were cumulatively injected with period of injection and shut-in it, is assumed that the surrounding rock has recovered its temperature during this shut-in day. Therefore the model was setup neglecting eventual change in the temperature distribution in the surrounding rock caused by earlier injection. Eleven straight injection days following this shut-in period were selected to test the models. The wellhead pressure and temperature and flow rates were very close. This made sure that the Joule-Thomson effect would be similar for all the day. The data were measured on an hourly basis. Therefore average daily pressure, and flow rate were considered for the 24 hours of each injection day. However the temperature of the last injection hour was considered as daily measured temperature. This is also justified by the fact the change in the surrounding temperature and consequently the effect of the heat transfer between the flowing fluid and the surrounding rock are function of the cumulative volume that flew in the wellbore.

Simulated and measured (at 1510 meters depth from the sea floor) values of pressure and temperature in the well are compared in table 7.6. Further comparisons are presented in Figure 7.6 and Figure 7.7 for pressure and temperature respectively.

The percentage of the deviation of simulated values from the measured ones has been calculated with equation 7.1.1. The highest percentages are 1.93 % for pressure and 10.83% for temperature.

Table 7.6: *Comparison of measured and simulated pressure and temperature in Snøhvit injection well*

Day	Pressure			Temperature		
	Bar Measured**	Bar Simulated	Deviation (%)	°C Measured**	°C Simulated	Deviation (%)
1	230.50	234.94	1.93	21.53	22.30	3.58
2	226.70	225.03	0.74	22.01	21.51	2.27
3	226.82	226.14	0.30	21.53	21.60	0.33
4	228.20	226.58	0.71	21.20	22.02	3.87
5	228.35	225.56	1.22	21.12	21.97	4.02
6	228.61	225.13	1.52	21.00	22.06	5.05
7	229.36	226.32	1.33	20.92	22.21	6.17
8	229.89	226.54	1.46	19.86	22.01	10.83
9	230.42	230.66	0.10	21.70	23.82	9.77
10	230.72	229.49	0.53	20.32	22.32	9.84
11	230.91	226.60	1.87	20.70	21.92	5.89

*** Measurement taken at 1510 meters*

Like in the Ketzin well the high accuracy in pressure calculation is justified firstly by the strict approach taken in well segmentation. The stringent validation criteria in the segments force the errors to be minimized. Secondly the high accuracy of the models in calculating not only the thermophysical properties (density, viscosity, etc.) of the fluid in the well but also in calculating the temperature of the fluid further contributes to minimize these errors. The uncertainties on surrounding rock thermal properties and to a certain extent on wellbore cement and tubular thermal properties are higher than the uncertainty on density and viscosity, used in calculating the pressure. Consequently there is higher deviation between simulated temperature values from measured values than between simulated pressure values from measured ones. However this deviation in temperature values is acceptable as it is generally below 10%.

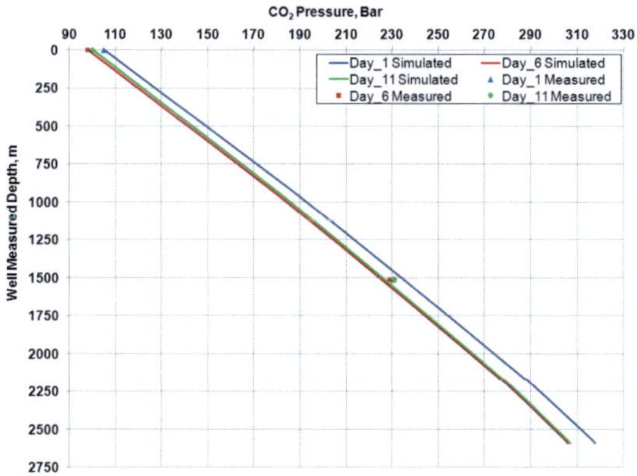

Figure 7.6: *Comparison of simulated pressures with measured pressure in Snøhvit injection well*

Figure 7.7 also shows the evolution of the rock-cement interface temperature with time. Initially (on day 1) the rock-cement interface temperature is assumed to be the undisturbed formation temperature and follows the geothermal gradient. As injection goes on, because the gas in the wellbore is colder than the surrounding rock, heat is transferred from the rock to the gas in the well. As a result the temperatures of surrounding rock and that at the rock-cement interface and the wellbore components decrease when that of the gas increases.

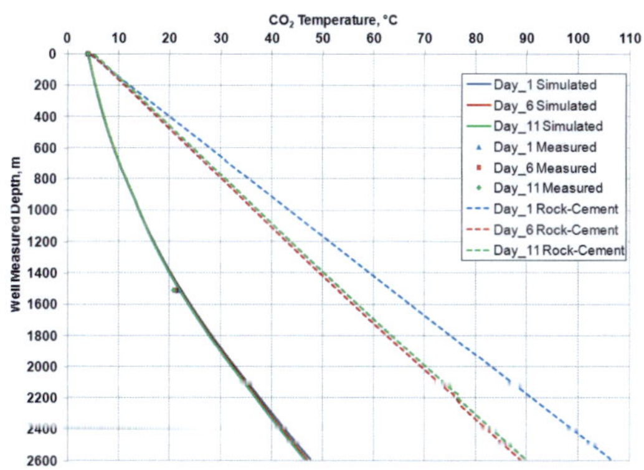

Figure 7.7: *Comparison of simulated temperatures with measured temperatures and prediction of the rock-cement interface temperature in Snøhvit injection well*

An example of how the model predicts the propagation of the temperature in the surrounding rock is shown in Figure 7.8. This figure also shows how the radius of thermal influence (RTI) at 1300 m depth is predicted to extend with time.

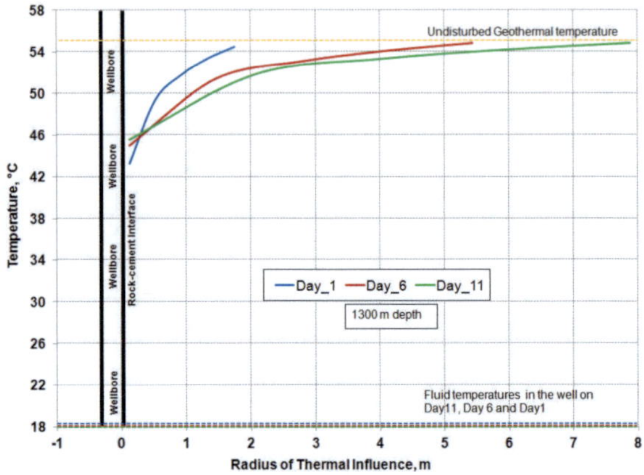

Figure 7.8: *Predictions of the evolution of rock temperature in the RTI at 1300 m depth in Snøhvit well*

The radius of thermal influence increases with time as injection goes on. Because the gas is colder than the surrounding rock, the rock gets colder. As a consequence, heat is transferred from the formation to the rock-cement interface and from the rock-cement interface to the gas through the tubular and cement. Like in the case of Ketzin, the extent of the radius of thermal influence with time is proportional to the thermal conductivity of the rock. It is worthy to note that the RTI of Snøhvit after 11 days of injection is smaller than that of Ketzin after only 8 injection days. This could be justified by the higher thermal conductivities considered in the Ketzin case (1.25 W/mK for Snøhvit and 3.0 and 2.5 and 1.75 W/mK for Ketzin). Although reasonable, it would not be true in all cases. Because, as will be discussed in the section dedicated to sensitivity analyses, the RTI does not depend only on thermal conductivity. Some other parameters such as the temperature of the injected fluid and the injection flow rate amongst others can also influence it.

With the above-presented injection cases it is shown that this work does not only make good predictions of temperature and pressure of CO_2 in an injection well but also that it can predict the evolution of the gas temperature as well as the propagation of temperature in the surrounding rock as a function of depth and time. Measurements in wells of different depths, inclination and different injection conditions have been reproduced within experimental uncertainty for pressure and temperature.

Nevertheless it would have been better if the predictions were matched to more than one point in the well. Unfortunately the data was measured at only one point in the injection wells. As earlier introduced measured data at several points along a well producing CO_2 of 97% purity have been presented by Cronshaw and Bolling (1982). This data is used in the next section to further test the capability of the models developed in this work.

7.2 Comparison with Measured Data from the Production Well

7.2.1 Input data

The production well has been presented by Cronshaw and Bolling (1982). Major input data consists of wellbore configuration data and the wellhead fluid variables (temperature, pressure and flow rate). For the purpose of simplification the tubing diameter used in the simulation is that of the longest section of the well. The wellbore configuration is summarized in table **7.7**. The produced gas is made up of 97% CO_2. The impurities have not been specified. The flow rates and bottomhole pressure and temperature as well as geothermal gradient are given in table 7.8. Rock and annulus thermal conductivity were taken as 1.5 W/mK and 0.4 W/mK respectively. Two cases have been simulated as presented in table 7.8. The results are presented in table 7.9 and in Figure 7.9 to Figure 7.12.

***Table 7.7**: Wellbore configuration and thermal conductivity data for the CO_2 production well*

Section	Item	Tubing	Casing	Annulus	Cement
1	**ID (m)**	0.0883	0.1594	0.102	0.1778
	OD (m)	0.102	0.1778	0.1594	0.2583
	Thickness (m)	0.007	0.0075	0.07	0.033
	TC (W/mK)	45	45	0.4	1.3
	Depth (m)	305	305	305	305
	Deviation (degrees)	0	0	0	0
2	**ID (m)**	0.0883	0.1594	0.102	0.1778
	OD (m)	0.102	0.1778	0.1594	0.2583
	Thickness (m)	0.007	0.0075	0.07	0.033
	TC (W/mK)	45	45	0.4	1.5
	Depth (m)	914	914	914	914
	Deviation (degrees)	26.5	26.5	26.5	26.5

***Table 7.8**: Input data for the CO_2 production well*

Cases	Flow Rate	Temperature		Geothermal Gradient	Bottomhole pressure
		Bottomhole	Surface		
	m^3/day	°C		°C/m	bar
1	115753	~45	15.56	~0.0328	88.6
2	342352	~45	15.56		87.3

7.2.2 Results and Discussion

As earlier introduced the pressure traverses are presented in Figure 7.9 and Figure 7.10 while temperature traverses are shown in Figure 7.11 and Figure 7.12. The data has been measured only for the first day of production. The results have been analyzed and summarized in table 7.9. This table shows that the pressure predicted with this work are very close to the measured data. This can also be seen on Figure 7.9 and Figure 7.10 for the first and second cases respectively. On some of the data points the deviation from the measured value is less than 0.5%. The maximum deviation is 1.56% and occurs in the second case where the flow rate is higher. This would be justified by a higher carryover of water in the gas stream. Water carryover is not accounted for by the models in this work.

***Table 7.9:** Comparison of measured and simulated pressure in the CO_2 production well*

MD (meters)	Pressure						Temperature					
	Measured Values		Simulated Values		Deviation (%)		Measured Values		Simulated Values		Deviation (%)	
	Case 1	Case 2	Case 1	Case 2	Case 1	Case 2	Case 1	Case 2	Case 1	Case 2	Case 1	Case 2
0	64	59.3	62.5	60	2.34	1.18	26.11	25	26.76	26.02	2.6	4.08
75	65	61	63	61.24	3.08	0.39	26.8	25.5	26.64	26	0.63	1.96
150	67	64	66	63	1.49	1.56	27.23	26.1	27.25	26.2	0.08	0.38
300	71	69	70.13	68	1.23	1.45	29.45	27.7	29.98	27.48	1.91	0.79
450	75	73	75.01	73.12	0.01	0.16	33.3	31.45	34.01	31.8	2.26	1.11
600	80	78.5	79	78	1.25	0.64	37.2	36.24	37.87	36.04	1.85	0.55
750	84	82	83.7	82.17	0.36	0.21	42.01	39.3	41.35	39.93	1.68	1.6
914**	88.6	87	88.6	87	0	0	45	44	45	44	0	0

*** This row contains initial conditions at the bottom of the well*

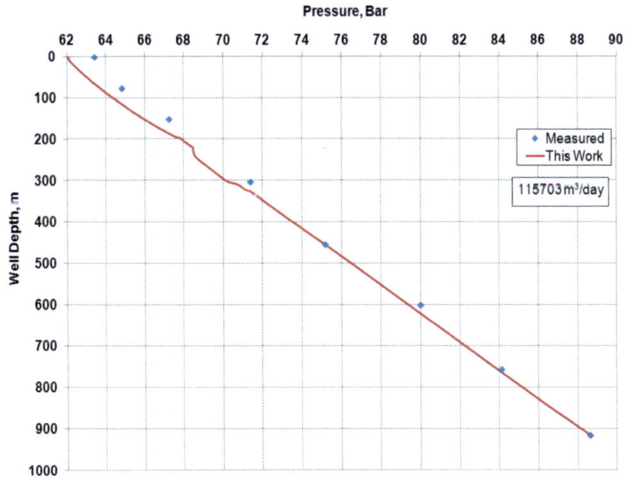

***Figure 7.9**: Comparison of simulated pressures with measured pressures along the production well for case 1*

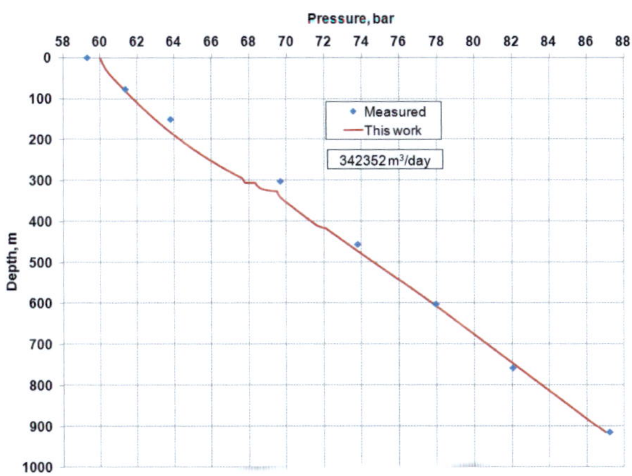

***Figure 7.10:** Comparison of simulated pressures with measured pressures along the production well for case 2*

The predicted temperature values are also very close to the measured ones as also shown in Figure 7.11 and Figure 7.12. The highest deviations occur at the wellhead and are also larger for the case with highest flow rate (case 2). This corroborates the above argument on water carryover by the gas stream. All the same the highest deviation is 4.08% which is still within experimental uncertainty.

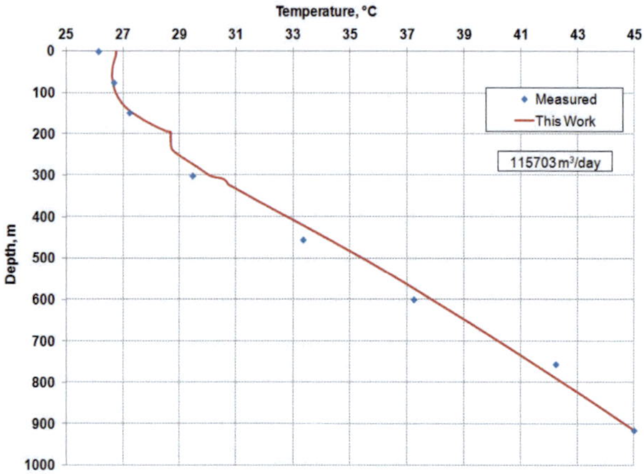

Figure 7.11: *Comparison of simulated temperatures with measured temperatures along well for case 1*

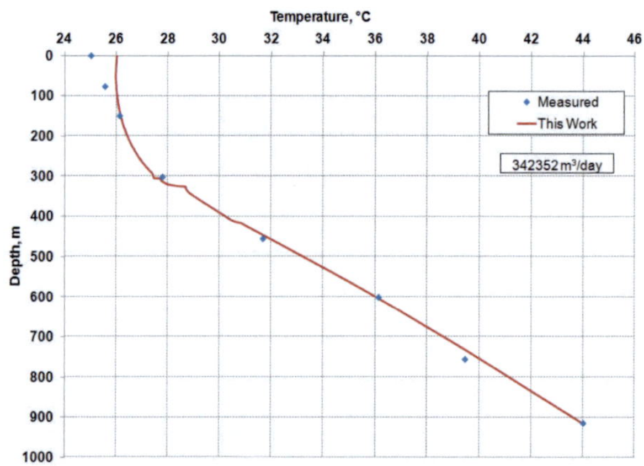

Figure 7.12: *Comparison of simulated temperatures with measured temperatures in the production well for case 2*

The foregone points help to conclude that this works reproduces measured production pressure and temperature data for the two cases quite satisfactorily.

Notwithstanding the good predictive performance of this work, it is important to critically look at the unexpected change in temperature gradient around the wellhead. In fact there is expectation of temperature decrease corresponding to decrease in pressure. However this work rather shows a slight increase in temperature with decreasing depth near the wellhead. Such increase becomes more important for the case with higher flow rate. This can be explained by the fact that as temperature decreases the liquid content of the fluid increases due to condensation. This is acknowledged in the codes of this work which. It considers that heat released during condensation reheats the fluid and increases its temperature. The same is expected in real life. However also in real flow Cronshaw and Bolling (1982) recognized that there is a very complex interaction between the frictional cooling\heating effects described in details by Katz (1972) and Christie (2007) and two phase flow which occurs around the wellhead in case 1 and case 2. The frictional cooling or heating effect is a little more complex when the fluid falls in the two-phase region. This is because the liquid phase will behave differently from the gas phase. Joule Thomson effect will be more pronounced in the gas phase than in the liquid phase because liquid has a higher density and a higher heat capacity than gas. This work models two-phase flow as a single phase with weighted average of properties of the vapour and liquid fractions, neglecting slip between the liquid and the vapour phase. As such effect of the Joule-Thomson cooling will be reduced by the presence of liquid. Additionally continuous condensation will release heat that goes to the formation to increase the temperature of the surrounding rock and result in heating of the fluid. With increasing liquid fraction, phase slip becomes very important and needs to be considered in two-phase flow modeling.

In real life phase slip causes liquid to drop down and contact the ascending fluid stream which is richer in gas. It is aggravated in this case by the presence of droplets of entrained reservoir water in the gas stream. The interaction between descending liquid and upcoming fluid will result in more condensation and evaporation in the two streams. This will cause the amount of heat released to the surrounding to decrease. Another consequence of phase slip is that the exiting stream will be richer in gas and thus the Joule-Thomson cooling effect will be more important. Therefore the temperature would continue decreasing notwithstanding the condensation and evaporation effects.

No efforts have been made to tune the model to enable it to account for the above-described phase slip and resulting evaporation and condensation processes. In fact these effects will be mitigated during injection. In event of two-phase flow, liquid and vapour will flow in the same direction towards the bottom of the hole. This reduces the phase slip and makes homogenous model to be more reasonable. Moreover as the well get deeper, CO_2 will fall in the supercritical region where it flows more effectively.

7.3 Comparison of the Different Temperature Models Developed in this Work

7.3.1 Performance Comparison

As earlier presented, four different temperature models have been developed within this work. The model used in the foregone simulations is the non-adiabatic temperature model. It takes into account both the temperature effects caused by compression/expansion and friction of gas the in the wellbore, and those caused by the exchange of heat between the surrounding rock and the flowing fluid. As such it is the main model to be used for simulations. The three other models are useful in studying the thermodynamic processes in the well in details by comparing their predictions with those of the main model as well as with each other. These are the adiabatic model, the geothermal model and the isothermal model.

As earlier presented the adiabatic model considers only the temperature effects caused by compression/expansion and friction of the gas in the wellbore. It is based on the assumption that the wellbore is insulated and as such the gas in the well is isolated from the surrounding. This situation would be approached at very high flow rates or in a near to perfectly insulated well.

The geothermal model assumes that the gas temperature has a linear temperature profile which is the same as the temperature of the rock surrounding the wellbore. This situation would be approached at very low flow rates.

In the isothermal model the gas temperature remains constant along the well. It assumes that the wellbore is insulated and that the temperature effects caused by compression or expansion of the gas in the well are negligible. This situation is approached in a horizontal well when frictional pressure losses are insignificant and when the inlet gas temperature equals the geothermal temperature.

The examination of the performance of these different models is done on a vertical well of 2000 meters long. The wellbore configuration and wellhead parameters are given in table 7.10. Two wellhead injection scenarios are considered and given in table 7.11. In the first scenario (scenario I) the CO_2 is injected in the liquid phase at 30°C and 60 bar which represent conditions close to those of the Ketzin well. The second scenario (Scenario II) represents conditions close to those of the Snøhvit well with temperature of 4°C and pressure 100 bar. As such the fluid is injected in the gas phase. Although CO_2

is injected as gas in scenario I and liquid in scenario II, gas, liquid and supercritical phases would be experienced along the well.

The flow rates are changed as a function of the models and processes investigated. However where not specified, the investigations are based on a flow rate of 200000 m^3/day.

Table 7.10*: Wellbore configuration and thermal conductivity data for the well used in comparisons*

Section	Item	Tubing	Casing	Annulus	Cement
1	ID (m)	0.076	0.1594	0.102	0.1778
	OD (m)	0.089	0.1778	0.1594	0.2583
	TC (W/mK)	45	45	0.4	1.4
	Measured Depth (m)	2000	2000	2000	2000

Table 7.11*: Input data for the well used in comparisons*

Scenarios	Flow Rate	Temperature			Geothermal Gradient	Wellhead Pressure
		Wellhead	Reservoir	Surface		
	m^3/day	°C			°C/m	bar
I	200000	30	70	10	~0.03	60
II	200000	4	70	10	~0.03	100

7.3.1.1 Results and Discussion

The pressures predicted by the four models are presented in Figure 7.13 for scenario I. In this figure, the predictions with geothermal and isothermal models result in a very large bottomhole pressure. This is as a result of the CO_2 changing from gas to liquid phase at the well head with the geothermal model, and somewhere along the well with the isothermal model. This can also be seen in Figure 7.14, which shows the phase of the gas in the well as a function of temperature and pressure predicted by the four models. The fluid remains strictly in gas and gas-like supercritical phase for the adiabatic and non-adiabatic models whereas for the isothermal model it transits early to the liquid phase and remains in this phase along the well.

In the geothermal model the gas flows as liquid in the upper part of the well and liquid-like supercritical fluid in the lower part. As a consequence the predictions of densities along the well are much higher for the isothermal and geothermal models than for the adiabatic and non-adiabatic models. This can be seen in Figure 7.15. In this figure the densities predicted by the isothermal model show a very smooth transition to the near-critical region and from the near-critical region to the liquid region. This is a consequence of the temperature remaining at 30°C which is very close to the critical temperature (31.04°C). The geothermal model goes immediately into liquid phase at the wellhead. This is as a result of sudden decrease of temperature from 30°C to 10°C, corresponding to surface temperature.

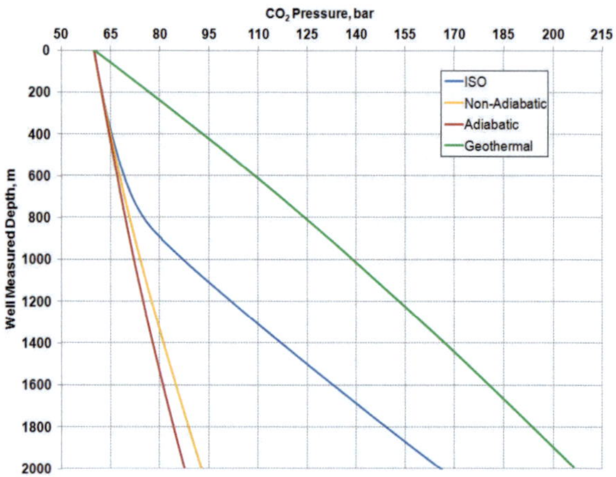

Figure 7.13: *comparison of the predictions of pressure along the well with the four different temperature models for wellhead injection scenario I*

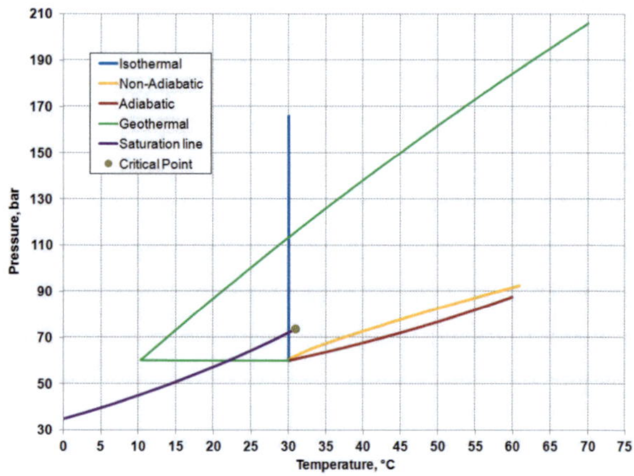

Figure 7.14: *comparison of the pressure predictions of the four models with the saturation line to show the phase of the gas in the well with wellhead injection scenario I*

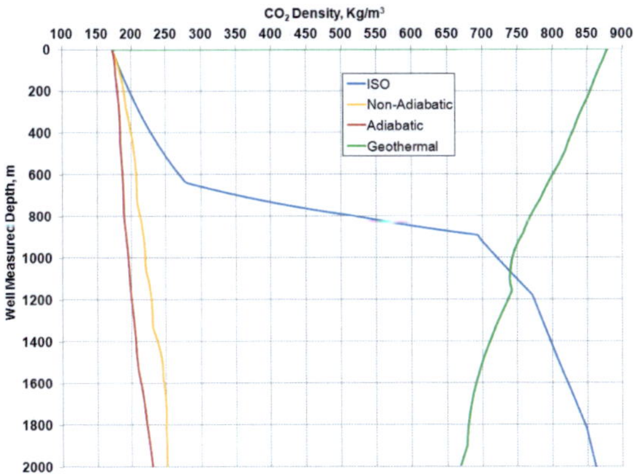

Figure 7.15: *Comparison of the predictions of densities along the well with the four different temperature models for the wellhead injection scenario I*

The predictions of the temperature along the wellbore are presented in Figure 7.16 below. In this figure the isothermal model predicts constant temperature at 30°C as expected and the geothermal model predicts the temperature to be equal to geothermal temperature. It goes from 10°C which is the surface to 70°C corresponding to surrounding rock temperature at the depth of 2000 meters. With the adiabatic model there is an increase in temperature as a result of increase in pressure along the well. This is caused by the Joule-Thomson effect and the heat of compression in the well. The same processes occur in the non-adiabatic case but in this case the surrounding rock also heats or cools the fluid resulting in a different temperature profile. For the current injection scenario (30 °C THT and 60 bar THP) the non-adiabatic model predicts initial decrease of temperature in the upper part of the well. This is a result of cooling by the surrounding rock which has a lower temperature (corresponding to geothermal temperature). The intensity of the cooling process reduces gradually as the temperature difference between the gas and the surrounding rock gets smaller. At a certain point around 1100 meters, the surrounding rock's temperature becomes higher than that of the gas. The rock heats up the gas from that point increasing the gas temperature towards the geothermal temperature.

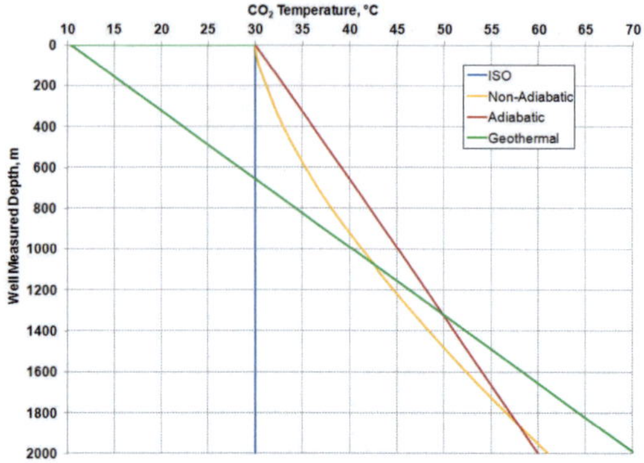

Figure 7.16: *Comparison of the predictions of temperature along the well with the four different temperature models for the gas phase injection scenario I*

For the liquid phase injection represented by scenario II, the isothermal and adiabatic models predict that CO_2 will flow strictly in the liquid phase, and the geothermal and non-adiabatic models predict that it will flow in the liquid and supercritical phases. This can be seen in Figure 7.17. As a result the difference of the predictions of pressure and densities by the four models is not as significant as in scenario I where the CO_2 is injected as gas. Figure 7.18 and Figure 7.19 show the pressure and densities predictions respectively. The geothermal model predicts the lowest BHP as a result of higher temperature predictions as presented in Figure 7.20.

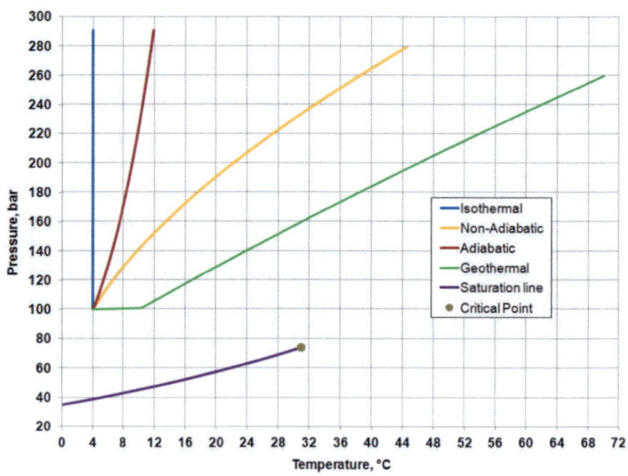

Figure 7.17: *Comparison of the pressure predictions of the four models with the saturation line to show the phase of the gas in the well with wellhead injection scenario II*

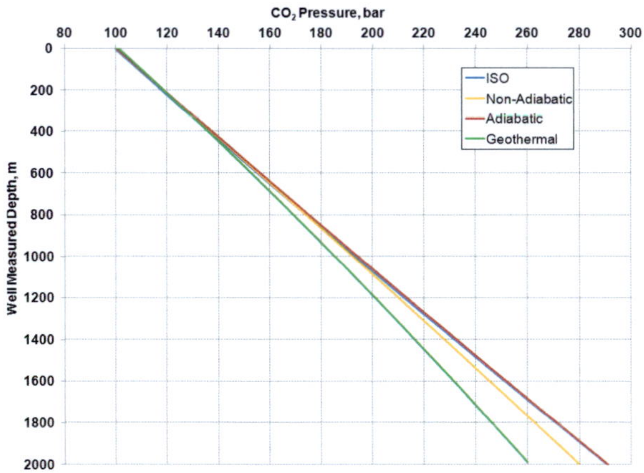

Figure 7.18: *Comparison of the predictions of pressure along the well with the four different temperature models for wellhead injection scenario II*

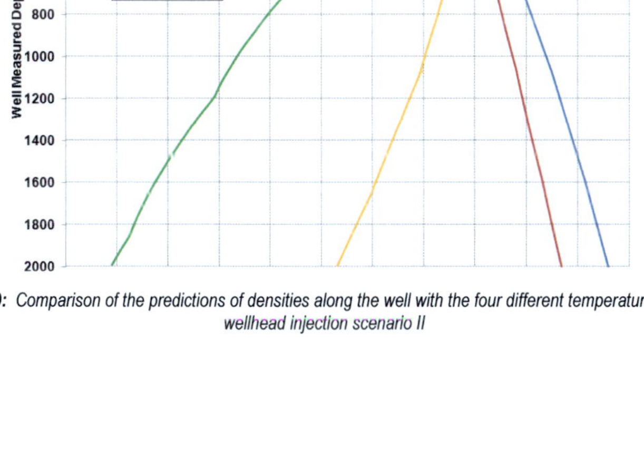

Figure 7.19: *Comparison of the predictions of densities along the well with the four different temperature models for wellhead injection scenario II*

Figure 7.20 also shows that the isothermal model predicts the temperature to remain constant at 4°C, equal to the wellhead temperature. Like in scenario I the geothermal temperature goes from 10°C corresponding to surface temperature to 70°C corresponding to the temperature of the surrounding rock at the depth of 2000 meters. However in this scenario the fluid is heated at the wellhead from 4°C to 10°C. It remains nevertheless in the liquid phase. The non-adiabatic model predicts that the gas is heated from the wellhead to the reservoir. Unlike in the first scenario, the wellhead temperature is smaller than the surface temperature. As such the surrounding rock temperature is higher than the fluid temperature throughout the well length. Similarly to scenario I the adiabatic model predicts that the fluid is heated in the wellbore as a result of compression. However the increase in temperature (~8°C) is much smaller than the increase in scenario I (~30°C). This is because the fluid flows in the gas and gas-like supercritical phases in scenario I whereas it flows strictly in the liquid phase in the current scenario. The Joule-Thomson coefficient is much higher for the gas phase than for the liquid phase. Figure 7.21 presents the Joule-Thomson coefficient predicted by the adiabatic model along the wellbore for the two injection scenarios. It is also worthy to mention that the heat of compression is also smaller in the liquid phase than in the gas phase. This is a consequence of the fact that liquid phase compressibility is very small compared to gas phase compressibility.

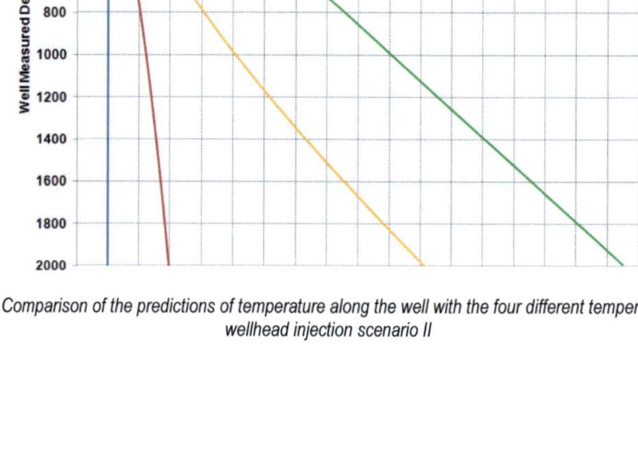

Figure 7.20: *Comparison of the predictions of temperature along the well with the four different temperature models for wellhead injection scenario II*

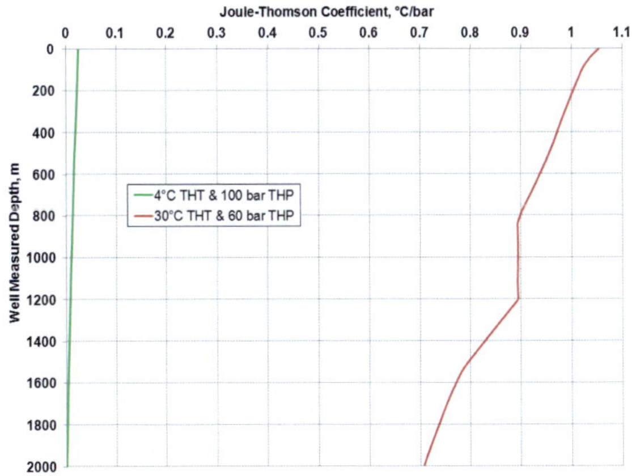

Figure 7.21: *Comparison of the predictions of the Joule-Thomson coefficient along the well for the two injection scenarios with the adiabatic model*

7.3.2 Similarities between the Main Temperature Model (Non-Adiabatic) and the other Temperature Models

At this point it would important to look at the similarities between the non-adiabatic temperature model and the other temperature models presented above. This will throw more light on the agreement of this work with thermodynamics and heat transfer theory.

7.3.2.1 Similarities of the Non-Adiabatic Model with the Geothermal Model

The temperature predictions of the main model will approach those of the geothermal model at low flow. This is due to the fact that the lower the flow rates the higher also the contact time between the fluid and the surrounding rock. Therefore the amount of heat that will be exchanged between the fluid and the rock will also be higher causing the fluid and rock temperature to approach each other much faster. It can be seen in Figure 7.22 and Figure 7.23 where the fluid temperature approaches the geothermal temperature with decreasing flow rates. Note that the temperature of the rock is considered to be almost constant. This would be the case only in the first days of injection. More insight in how the two temperatures will approach themselves while changing with injection time is given in the section on sensitivity analyses.

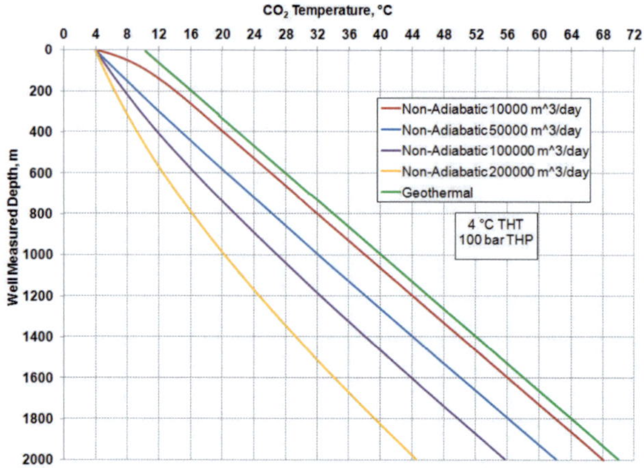

Figure 7.22: *Comparison of the predictions of temperature by the non-adiabatic model with the prediction of the geothermal model as a function of flow rate (for wellhead injection scenario II)*

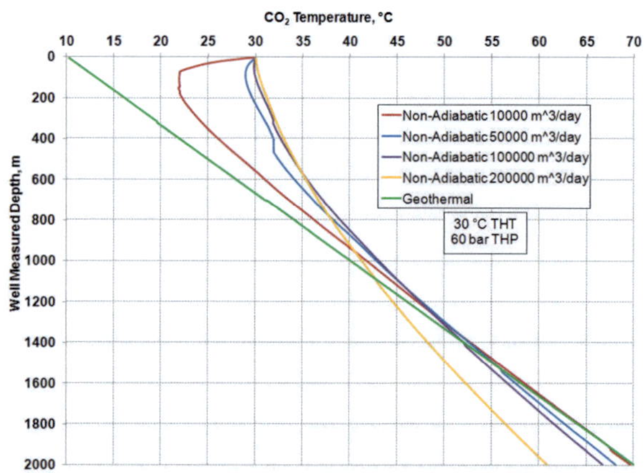

Figure 7.23: *Comparison of the predictions of temperature by the non-adiabatic model with the prediction of the geothermal model as a function of flow rate (for wellhead injection scenario I)*

As earlier discussed in chapter five the rate of heat transferred between the surrounding rock and the fluid is controlled by the dimensionless number $\aleph$. Equation 7.3.1 below recalls how the rate of heat transferred between the fluid and the surrounding rock ($\dot{q}$) is calculated in the codes.

$$\dot{q} = c_p \dot{m} \aleph \Delta T_{fF} \tag{7.3.1}$$

Where $\dot{m}$ is the mass flow rate, c_p is the specific heat capacity at constant pressure and ΔT_{fF} is the temperature difference between the fluid and the surrounding rock. $\aleph$ is a function of the overall heat transfer coefficient (U_{over}) and the heat transfer area (A_{over}), the fluid specific heat capacity at constant pressure and the mass flow rate as also recalled in equation 7.3.2 below. All other parameters being equal, the larger the flow rate the smaller the value of Niu and the lesser the heat transferred between the fluid and the surrounding rock.

$$\aleph = \frac{U_{over} A_{over}}{c_p \dot{m}} \tag{7.3.2}$$

Table 7.12 gives the different values of $\aleph$ and other parameters on which it is dependent, for the different flow rates presented in Figure 7.22 representing for wellhead injection scenario II. The higher values of heat capacities correspond to lower flow rates and thus higher temperatures. It is in agreement with the fact that heat capacity increases with temperature. Although it also increases with density the increase with density is not as high as the increase with temperature. The overall heat transfer coefficient (OHTC) is proportional to density and viscosity. As such higher densities and viscosities will yield higher OHTC. However the major controlling parameter for $\aleph$ is the flow rate.

***Table 7.12**: Predicted thermophysical and heat transfer parameters with non-adiabatic model (first scenario)*

	Geothermal	**10000 m³/day**	**50000 m³/day**	**100000 m³/day**	**200000 m³/day**
Cp (KJ/Kg.K)	52.7	52.32	50.3	49.3	47.8
OHTC(W/m².K)	N/A**	16.18	21.7	23.73	25.24
Density (Kg/m³)	837	845	878	901	926
Viscosity (Cp)	0.078	0.08	0.088	0.093	0.099
$\aleph$	N/A	0.089	0.021	0.009	0.0032

** The Geothermal model does not allow for calculation of OHT or $\aleph$

It is also worthy to note in Figure 7.22 and Figure 7.23 that the longer the well, the closer the non-adiabatic temperature will be to the geothermal temperature. This corroborates the argument that the longer contact time between flowing fluid and surrounding rock will reduce the temperature difference between the fluid and the surrounding rock.

7.3.2.2 Similarities of the Non-Adiabatic Model with the Adiabatic Model

The temperature predictions of the main model will approach those of the adiabatic model at higher flow rates. The higher the flow rate the smaller the contact time with the surrounding rock and therefore the smaller the amount of heat exchanged between fluid and rock. Figure 7.24 and more obviously Figure 7.25 illustrate this point. The non-adiabatic temperatures are closer to the adiabatic temperature in Figure 7.25 because the fluid is in the gas phase whereas it is in the liquid phase in Figure 7.24. As such in the predictions in Figure 7.25 the heating of the gas due to compression is more pronounce than the heating/cooling due to interaction with the surrounding rock as compared with the prediction in Figure 7.24. It is important to mention that the tubing diameters have been chosen so that the frictional pressure losses are almost equal in the four cases so that the temperature drops due to friction are also similar.

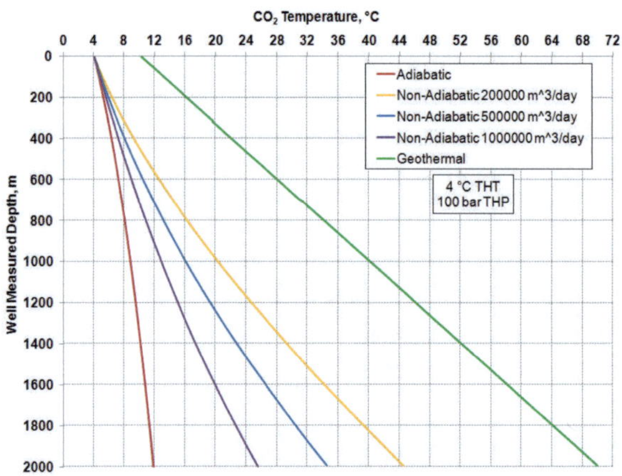

Figure 7.24: *Comparison of the predictions of temperature by the non-adiabatic model with the prediction of the adiabatic model as a function of flow rate (for wellhead injection scenario II)*

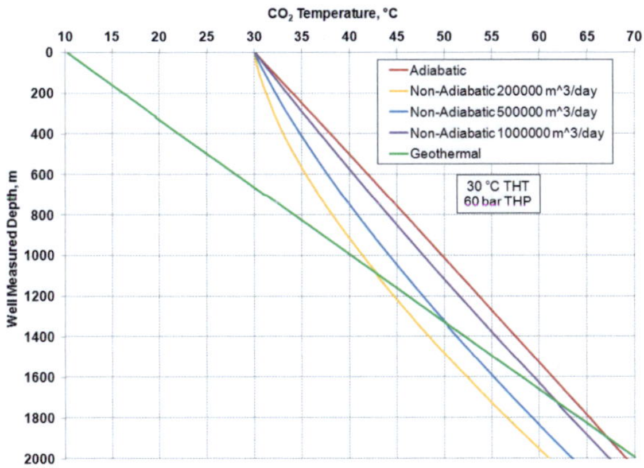

Figure 7.25: *Comparison of the predictions of temperature by the non-adiabatic model with the prediction of the adiabatic model as a function of flow rate (for wellhead injection scenario I)*

The different influence that compression/expansion and friction have on the fluid's temperature depending on the phase of the fluid in the well can be better illustrated by looking at the two earlier-defined thermodynamic parameters Psi (φ) and Omega (ω). Psi controls the temperature change due to compression/expansion while Omega controls the temperature drop caused by friction. This would be clearly illustrated by recalling the expressions of Psi and Omega in equation 7.3.3 and equation 7.3.4 below. The way they control the temperature in the well has been shown in equation 4.4.25.

$$\varphi = \eta \frac{P}{T} + z \frac{R}{C_{pm}} \tag{7.3.3}$$

$$\omega = \eta \frac{T}{P} \tag{7.3.4}$$

Where,

z is the compressibility factor

R is the universal gas constant

C_{p_m} is the isobaric molar heat capacity

η is the Joule-Thomson coefficient

P is the pressure

T is the temperature

In equation 7.3.3 it appears that Psi is directly proportional to the Joule-Thomson coefficient and the z-factor. As such the larger values of Psi would occur in the gas phase because as earlier discussed; gas phase has higher Joule-Thomson coefficient and higher z-factors than other phases. Similarly in equation 7.3.4, the value of Omega will be higher in the gas phase than in the other phases because of the direct proportionality of Omega to the Joule-Thomson coefficient. It is noteworthy that the value of Omega would generally be higher than that of Psi but its influence on temperature will not be as important as that of Psi. This is because Omega controls the temperature drop due to frictional pressure losses and Phi controls the temperature change due to compression/expansion of the gas in the well. Frictional pressure loss is generally very small, compared to the pressure increased or lost due to compression or expansion in the well. Figure 7.26 and Figure 7.27 show the predicted values of Phi and Omega by the adiabatic model along the well for scenario I and scenario II, where the fluid flows in the gas and liquid phases respectively. It is also important to note that at the same temperature the liquid phase normally has higher pressures and very small z-factors. These also explain why Psi and Omega would be smaller in the liquid phase than in the gas phase.

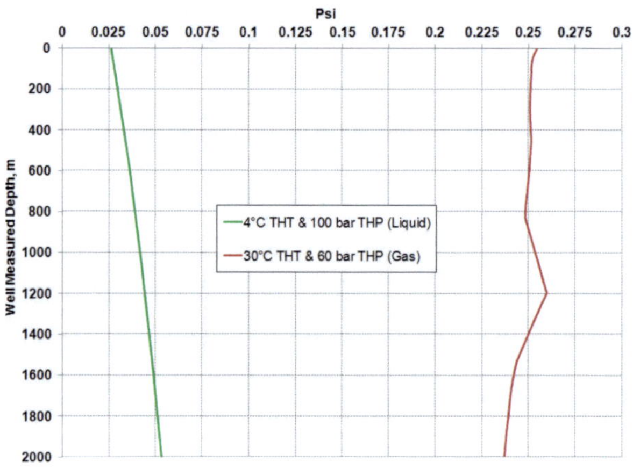

Figure 7.26: *Predictions of Psi by the adiabatic model as a function of the phase of the fluid in the well*

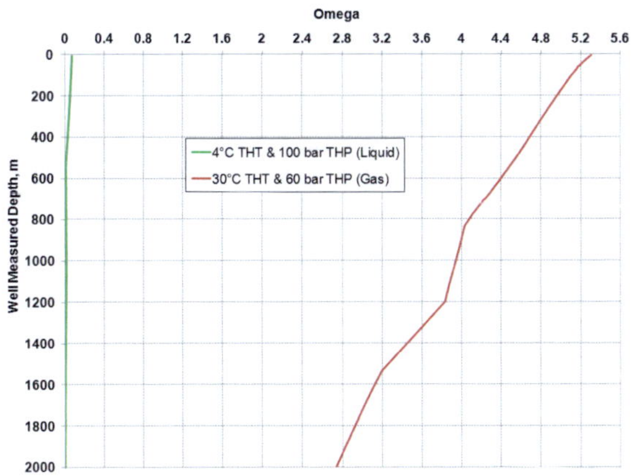

Figure 7.27: *Predictions of Omega by the adiabatic model as a function of the phase of the fluid in the well*

7.3.2.3 Similarities of the Non-Adiabatic Model with the Isothermal Model

The temperature predictions of the main model will approach those of the isothermal model in a horizontal well when injection temperature is equal to surrounding rock temperature provided that frictional pressure losses are small enough to not to induce significant cooling. Figure 7.28 shows how the temperature predictions are close to each other. As earlier mentioned it is worthy to note that the phase of the gas in the well also determines how the predicted non-adiabatic temperature will be closer to the predicted of the isothermal temperature. Temperature change due to expansion/compression is more important in gas flow than in liquid flow. In a horizontal well the pressure change resulting in a decrease in temperature is induced only by frictional losses. This is why the temperature predicted for gas flow is smaller than the one predicted for liquid flow in Figure 7.28. The difference although small can still be reduced by using tubing of larger inner diameter or of smaller roughness of the inner surface. These would help reduce the frictional pressure losses which are normally more important in gas flow. Insight on sensitivity of the pressure and temperature predictions to the tubing inner diameter and surface roughness will be given in the next sections.

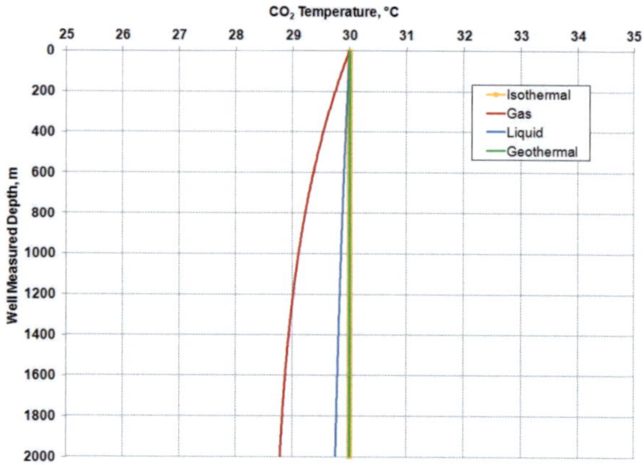

Figure 7.28: *Comparison of the predictions of temperature by the non-adiabatic model prediction of the isothermal model*

7.4 Analysis of Sensitivity of the Predictions of the Non-Adiabatic Model to Major Parameters that Would Influence Pressure and Temperature Prediction in the Well

The simulations are run on a vertical well having configuration given in table 7.10. The results are presented and discussed in the following sections. The rock thermal properties are the same as summarized in table 7.11.

7.4.1 Sensitivity of Pressure and Temperature to Wellhead Temperature

The sensitivity to wellhead temperatures aims at analyzing how the model predicts the evolution of temperature and pressure of the fluid along the well as a function of the wellhead temperature. Three wellhead temperature scenarios are considered: 20°C, 40°C and 60°C. The wellhead pressure is 50 bar for the three scenarios and the flow rate is 100000 m^3/day.

7.4.1.1 Results and Discussion

Figure 7.29 presents the pressure traverse in wellbore at 20°C, 40 °C and 60°C wellhead temperatures. It can be seen that the higher the wellhead temperature the lower the pressure of the fluid along the well. Although the fluid flows strictly in the gas phase in the three scenarios, the scenarios with higher wellhead temperature (40 and 60 °C) show almost similar pressure trend which deviate significantly from the one with the 20°C. This is justified by the fact that at 20°C and 50 bar, CO_2 is near its saturation region. This occurs in the upper part of the well. Therefore the densities are higher and very sensitive to temperature variations. Conversely, in the cases with 40°C and 60°C wellhead temperatures, at 50 bar wellhead pressure, the fluid is far from the saturation region in the upper part of the well. It therefore has lower densities which are also not very sensitive to temperature variations. This justifies lower hydrostatic pressures corresponding to higher temperatures (see figure C3 in Appendix C).

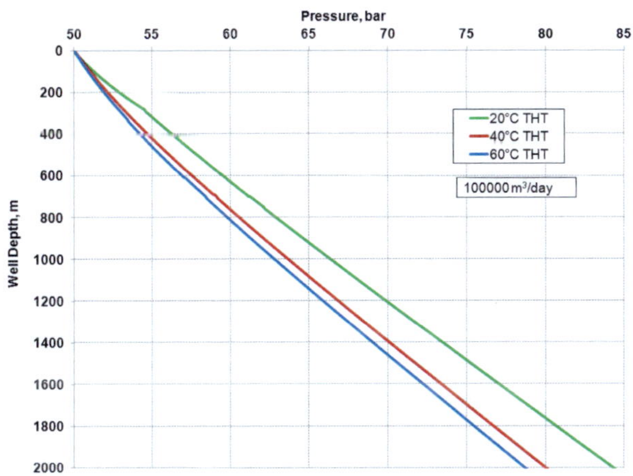

Figure 7.29: *Sensitivity of pressure to wellhead temperature*

The predictions of the densities of the fluid along the well can be seen in Figure 7.30. The scenario with 20°C shows a sharp hike at around 200 meters depth. This corresponds to saturation line where densities are calculated in this work as the average of liquid and vapour densities.

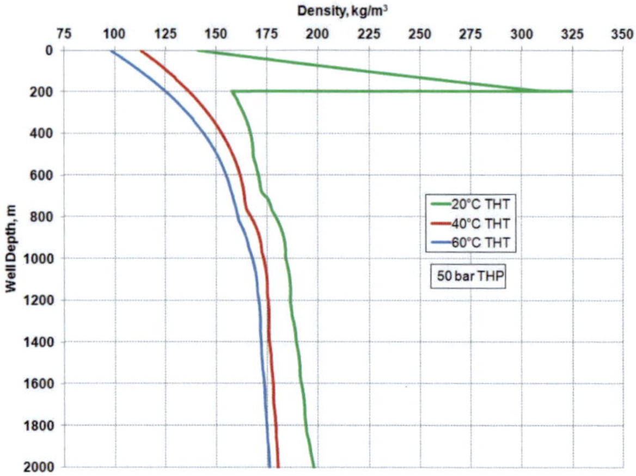

Figure 7.30: *Sensitivity of density to wellhead temperature*

The difference of the bottomhole pressures and the pressures in general amongst the three scenarios is less than 10 bar. Although obvious, this difference is not as high as it would be expected in light of the difference between the densities of the fluid at the wellhead. This is because there is fast cooling of the fluid in the upper part of the well, where its temperature is higher than surrounding temperature. As such the density difference diminishes as the fluid descends in the well.

The rate of cooling is proportional to the temperature difference between the fluid and the surrounding. Therefore the temperatures of the fluid in the three cases will approach each other as the fluid descends, and eventually all converge towards the surrounding rock's temperature in the lower part of the well. This can be seen in Figure 7.31, where the three scenarios result in almost equal bottomhole temperatures. The bottomhole temperatures get closer to that of the surrounding rock, which plays the role of regulator. In its regulator's role, the surrounding rock will cool the fluid whenever the fluid temperature is higher than its temperature and heats it up whenever the fluid temperature is lower than its temperature. In all the cases the fluid's temperature will eventually converge towards surrounding rock's temperature with increasing depth and time.

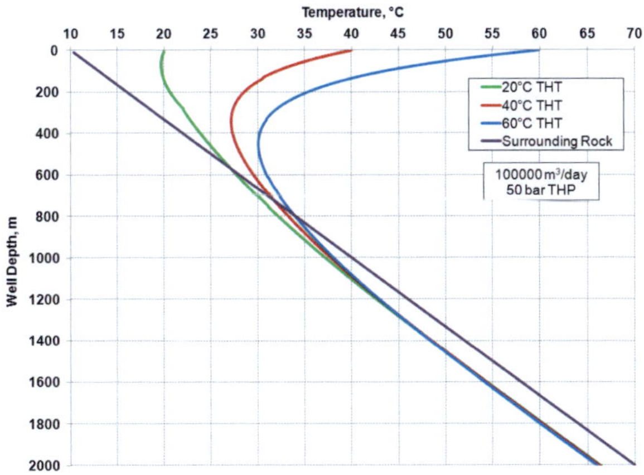

Figure 7.31: *Sensitivity of temperature to wellhead temperature*

In conclusion, it is worthy to highlight that at equal flow rate the bottomhole temperature as predicted by this work is almost not sensitive to wellhead temperature because of the interaction with the surrounding the surrounding rock. However different flow rates or different injection phase at the wellhead would result in different temperature profile and increase sensitivity of bottomhole temperature to wellhead temperature. This reflects not only the dependency of the interactions with the surrounding rock on flow rate but also the fact that compression has a higher effect on temperature in the gas phase than in the liquid phase. Sensitivity of bottomhole temperature to flow rate is presented in section 7.4.3.

7.4.2 Sensitivity of Pressure and Temperature Wellhead Pressure

The sensitivity to wellhead pressures aims at analyzing how the model predicts the evolution of temperature and pressure of the fluid along the well as a function of the wellhead pressure. Three wellhead pressure scenarios are considered: 40 bar, 60 bar and 80 bar. Injection takes place at 20°C Wellhead temperature and at a flow rate of 100000m³/day. The results are presented and discussed in the following section.

7.4.2.1 Results and Discussion

Figure 7.32 shows that the wellhead pressure has a great influence on the pressure in the well. In fact the pressure determines the phase and the density of the fluid. At 60 bar and 80 bar wellhead pressures the fluid is in the liquid phase at the wellhead and in the upper part of well before entering in the supercritical region in the remaining part of the well. Conversely, it is in the gas phase at 40 bar wellhead pressure before eventually entering the supercritical region. Because the liquid densities are by far larger than gas densities, the scenarios with 60 and 80 bar wellhead pressures will result in by far higher pressures in the well compared with the 40 bar wellhead pressure scenario. Although in all the scenarios, the fluid may enter the supercritical region at the lower part of the well, the 60 and 80 bar wellhead pressure scenarios are in the liquid-like supercritical phase, which is still of larger density than the gas-like supercritical phase in which the fluid flows in the scenario with 40 bar wellhead pressure.

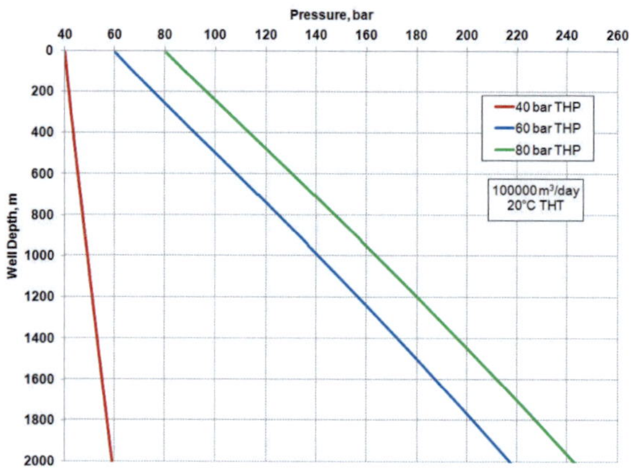

Figure 7.32: *Sensitivity of pressures to wellhead pressure*

The higher pressures do not only cause higher hydrostatic head (see figure C4 in Appendix C), but they also result in lower frictional losses in the gas phase flow, as a consequence of higher gas velocity. All these contribute to the prediction of large difference between bottomhole pressures of the scenarios with liquid phase injection (60 and 80 bar) and the bottomhole pressure of the scenario with gas phase injection (40 bar) in Figure 7.32. It is also worthy to note that the difference between the bottomhole pressures for the 60 and 80 bar injection is higher than the difference between wellhead pressures (20 bar) although both start in liquid phase. This re-emphasizes the influence of pressure on fluid density. In fact at 80 bar the fluid is denser and therefore yields a higher hydrostatic head.

Contrarily to the fluid pressure in the well, the fluid temperature is not too sensitive to the wellhead pressures, especially when the phase of the fluid does not change along the well. In Figure 7.33 it can be seen that the temperature difference reflects the fluid phase along the well (60 and 80 bar wellhead pressure scenario have same profile). The difference is related to the difference in Joule-Thomson coefficients and compressibility of the fluid amongst the phases. As earlier mentioned, in the gas phase the fluid has a higher Joule-Thomson coefficient and compressibility than in the liquid phase. This justifies the higher increase in temperature as a result of compression in the gas phase. Therefore the scenario with 40 bar wellhead pressure predicts higher bottomhole temperature because the fluid remains in the gas phase in the major part of the well. The Difference in Joule-Thomson coefficients amongst the phases is presented in figure C5 of Appendix C.

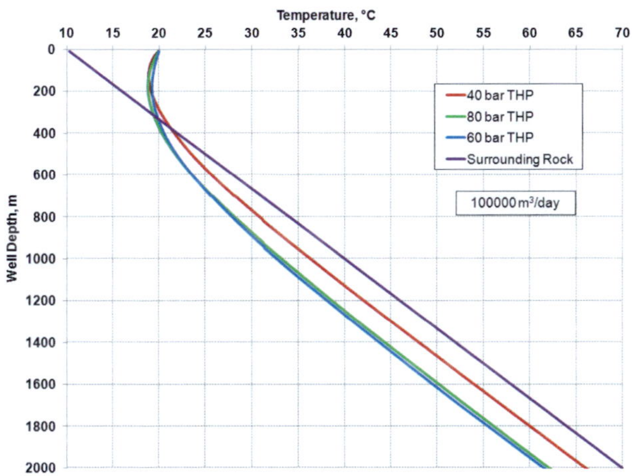

Figure 7.33: *Sensitivity of temperature to wellhead pressure*

In conclusion, it can be said in the light of the foregoing that the model in the present work predicts that higher tubing head pressure will results in significantly higher bottomhole pressure. This is a result of higher hydrostatic head and a lower frictional pressure loss caused by higher densities and is in agreement with the physics of fluid flow.

The influence of tubing head pressure on bottomhole temperature will not be very significant as it will be on the bottomhole pressure. However large difference in tubing head pressure would result in some difference in bottomhole temperature. This difference is due to the fact that at higher pressures (resulting from higher wellhead pressure) the fluid has lower Joule-Thomson coefficient and compressibility. Therefore smaller increase in temperature will be induced by same increase in pressure. However the temperature difference will be attenuated by the influence of the surrounding rock, which will cool the fluid whenever the fluid's temperature is higher than rock's temperature and will heat it up whenever its temperature is lower than rock's temperature.

7.4.3 Sensitivity of Pressure and Temperature to Flow Rate

The sensitivity to flow-rate aims at analyzing how this work predicts the evolution of temperature and pressure of the fluid along the well as a function of the flow-rates. Three scenarios are considered in this effect: 100000 m^3/day, 175000 m^3/day and 250000 m^3/day flow rates. The results are presented and discussed in the following sections.

7.4.3.1 Results and Discussion

Figure 7.34 presents the prediction of the evolution of pressure in the wellbore as a function of the flow rate. It can be seen that higher flow rates result in lower pressures along the wellbore and consequently in a lower bottomhole pressure. This is explained by the fact that the higher the flow rate the higher the frictional pressure losses and the lower the hydrostatic pressure head. Figure 7.35 and Figure 7.36 illustrate this point better by showing the profile of the hydrostatic and frictional pressure along the well as a function of the flow rates.

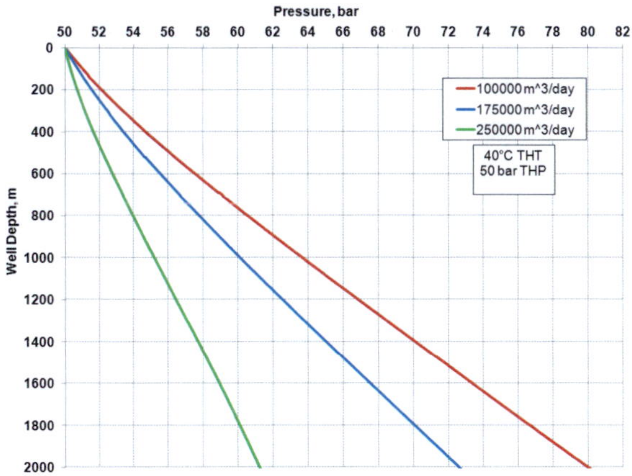

Figure 7.34: *Sensitivity of pressure to flow rate*

It would important to remember at this point that the relationship between the pressure in the well and the hydrostatic and frictional pressures is modeled in this work through the gravity number (N_g) and friction number (N_f) recalled in equation 7.3.5 and equation 7.3.6 below.

$$N_g = \frac{M_m L}{z_m R T_{in}} g cos\alpha \qquad 7.3.5$$

$$N_f = \frac{2 z_m f_f \dot{m}^2 R L T_{in}}{D A^2 M_m P_{in}^2} \qquad 7.3.6$$

Where,

M_m is the molecular weight, L is the segment's length, z_m is the Z-factor, R is the universal gas constant, T_{in} is the temperature at the inlet of the segment, g is the acceleration due to gravity, α is the deviation angle of the well segment, f_f is the fanning friction factor, A is the cross sectional area of the tubing, D is the diameter of the tubing and P_{in} is pressure at the inlet of the well segment.

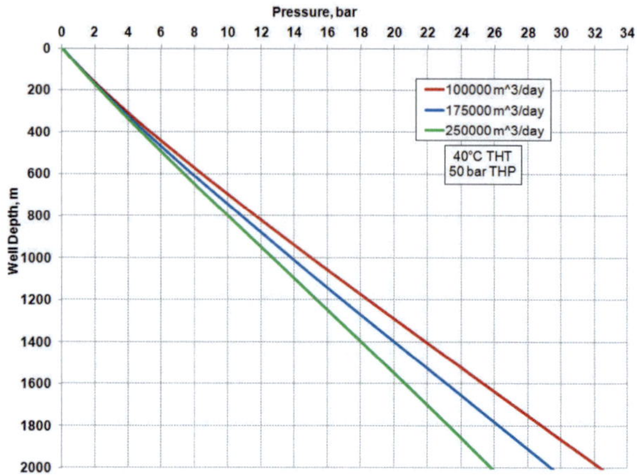

Figure 7.35: *Sensitivity of hydrostatic pressure to flow rate as a result of the pressure in the well shown in Figure 34*

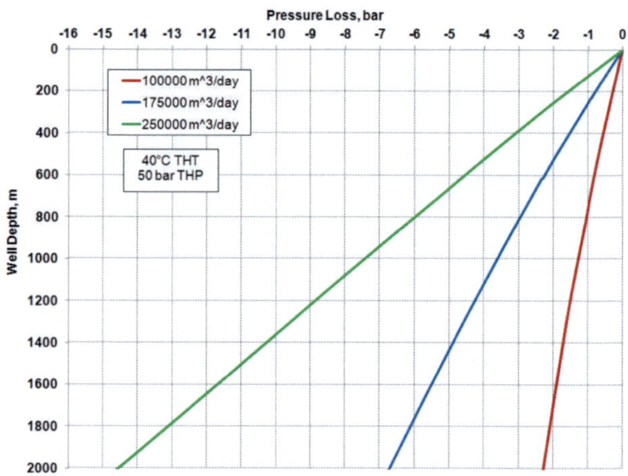

Figure 7.36: *Sensitivity of frictional pressure loss to flow rate*

As earlier explained, N_g gives an indication of the percentage of hydrostatic pressure contribution to the total pressure in the well segment. N_f informs on the magnitude of frictional forces in the well segment compared to the pressure forces at the inlet of the segment. Figure 7.37 and Figure 7.38 show how N_g and N_f are predicted to evolve in the well as a function of flow rate.

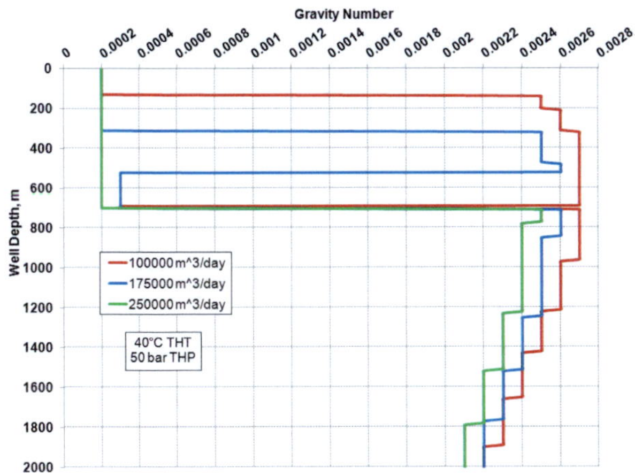

Figure 7.37: *Sensitivity of gravity number to flow rate*

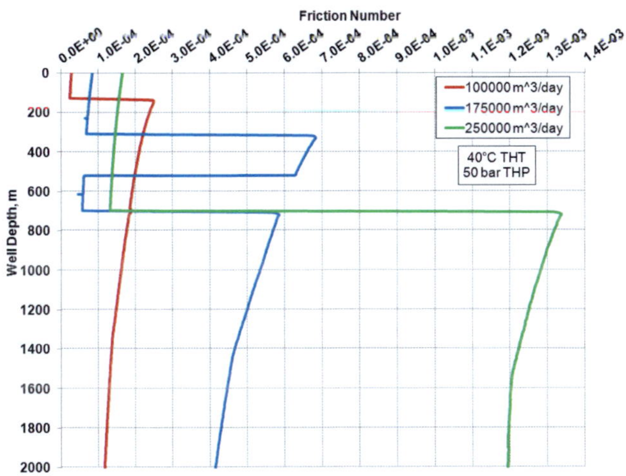

Figure 7.38: *Sensitivity of the friction number to flow rate*

It is worthy to note that N_g and N_f are also functions of the segment's length. The segment's length is automatically set as a function of the pressure and temperature validation criterion in the well. To recall it, this criterion entails that the relative difference between the z-factors at the inlet and outlet of the segment should not exceed 0.1%. This justifies the fact that the trend of N_g and N_f is changing along the well. However in essence higher flow rates result in smaller N_g because they induce higher frictional losses which result in smaller pressure and hence smaller densities which in turn will result in smaller hydrostatic head. Conversely higher flow rates generally result in higher N_f as a consequence of the higher frictional pressure losses that they would induce.

It is important to acknowledge that the magnitude of N_f also depends on the flow conduit inner diameter, the fluid flow-rate and the flow conduit inner roughness. In this sense, conduit inner diameter and roughness will be chosen so as to minimize frictional losses. Therefore it would be necessary to analyze how this dependence is predicted in this work. This is done in the section dedicated to sensitivity of pressure and temperature to the inner diameter and inner surface roughness of the tubing string.

The difference amongst the predicted wellbore temperatures is also quite important and is also in line with thermodynamics and heat transfer theory. As shown in Figure 7.39 at lower flow rates the fluid temperature approaches the surrounding rock temperature faster. In fact at higher flow rate, the fluid has smaller contact time with the well and therefore the heating and/or cooling effect of the surrounding rock is also smaller. This fact has earlier been discussed in relation with the dimensionless number $\aleph$ of which the value approaches zero with increasing flow rates. As a consequence, the temperature of the fluid will more quickly approach the surrounding rock temperature at lower flow rates. This is valid both for cooling (as in the upper part of Figure 7.39) and for hating (in the lower part of Figure 7.39) in the well.

In conclusion it can be said that the predictions of the influence of the flow rate on the bottomhole temperature and pressure are in line with the physics of fluid flow in pipes as well as with thermodynamics and heat transfer theory. As such higher flow rates will result in lower bottomhole pressure and temperature.

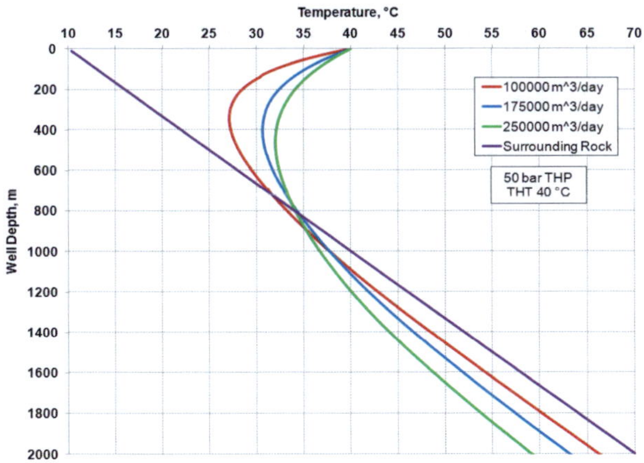

Figure 7.39: *Sensitivity of temperature to flow rate*

7.4.4 Sensitivity to Tubing Inner Diameter and Tubing Inner Surface Roughness

The sensitivity to tubing inner diameter and surface roughness aims at analyzing how the model predicts the evolution of temperature and pressure of the fluid along the well as a function of these parameters. Three diameters and surface roughness scenarios are considered in table 7.13 and table 7.14. The base case scenarios are those used in the foregone simulations and the other scenarios are for sensitivity investigations. The flow rate is set at 200000 m³/day in order to allow the fluid to flow in tubing of small diameter or large roughness and to enable the effects of frictions to be sensitive in larger tubing diameters or smaller roughness at the same time. The wellhead temperature and pressure are set at 50°C and 80 bar respectively so that phase transition would be avoided in the well. The fluid will flow only in the gas-like and (at the top) and liquid-like (at the bottom) supercritical phases. Wellbore configuration and other parameters are the same as in table 7.10 above.

***Table 7.13**: Summary of the scenarios used in investigating the sensitivity to tubing inner diameter*

Scenarios	Tubing			Flow Rate
	OD	ID	Inner roughness	
	meters (inches)	meters (inches)	meters	m³/day
I (base case)	0.089 (3.5")	0.076 (2.992")	0.00005	200000
II	0.073 (2.875")	0.0574 (2.259")		
III	0.1016 (4.0")	0.0883 (3.476)		

***Table 7.14**: Summary of the scenarios used in investigating the sensitivity to tubing inner roughness*

Scenarios	Tubing OD meters (inches)	Tubing ID meters (inches)	Tubing Inner roughness meters	Flow Rate m^3/day
I (base case)	0.089 (3.5")	0.076 (2.992")	0.00005	200000
II			0.0005	
III			0.005	

7.4.4.1 Results and Discussion

Figure 7.40 shows the profile of CO_2 pressure in the well. Smaller pressures are predicted for the tubing with smaller diameter and larger pressures for the tubing with larger diameter. However the difference between the pressures predicted by the two larger diameters is smaller compared with that predicted for the smallest diameter. It would imply that for a certain flow rate there is a critical tubing size above which the effect of tubing diameter on pressure does not change significantly.

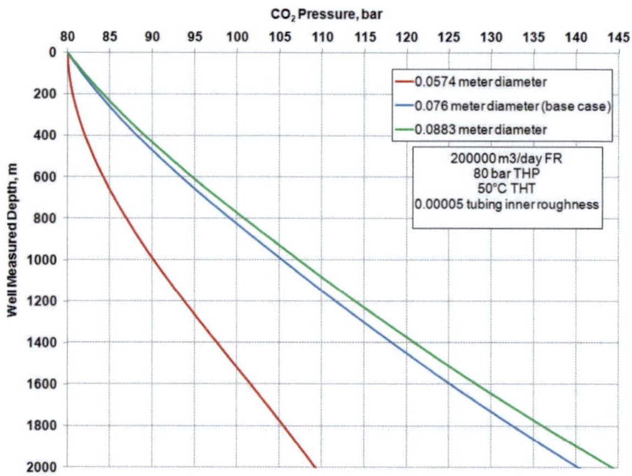

***Figure 7.40:** Sensitivity of pressure to tubing inner diameter*

The predictions of pressure profiles with different tubing inner surface roughness are presented in Figure 7.41. Smaller pressures are predicted for tubing of larger inner surface roughness. Noteworthy is that, like with the tubing inner diameter there is a certain critical roughness. Below this roughness the effect of roughness on pressure will become negligible. The critical tubing inner surface roughness and critical tubing diameter will not be discussed here because the focus is rather on the physical explanation of the changing trends of the pressure profiles with respect to these parameters and their consequence on the predictions of temperature profile in the well. In general, the larger the tubing inner diameter the smaller the frictional pressure losses. Conversely the smaller the tubing inner surface roughness the smaller the frictional pressure losses.

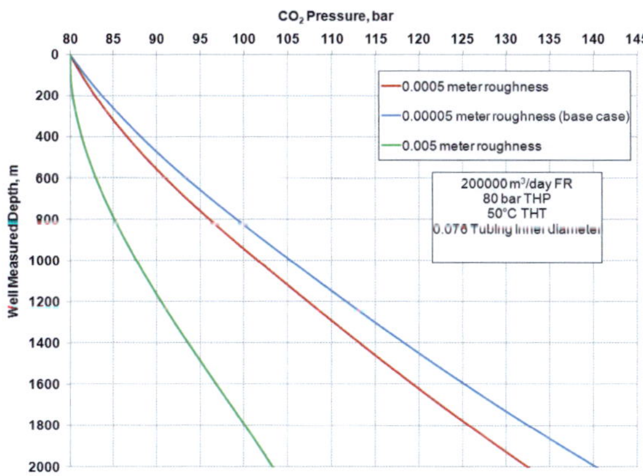

Figure 7.41: *Sensitivity of pressure to tubing inner roughness*

Figure 7.42 and Figure 7.43 show the predictions of the frictional pressure losses along the well as a function of tubing inner diameter and tubing inner surface roughness respectively. Smaller diameters result in larger frictional pressure losses and in contrary smaller roughness will yield smaller frictional pressure losses. This would be obvious when inspecting the friction number in equation 7.3.6. The higher the friction number the higher the frictional losses. The friction number is inversely proportional to the tubing inner diameter and directly proportional to the friction factor. The friction factor is also directly proportional to tubing inner roughness but inversely proportional to tubing inner diameter. These explain why larger pressures are predicted for larger tubing inner diameter and lower pressures are predicted for larger tubing inner roughness.

Insight can also be obtained on how the effects of the tubing inner diameter and tubing inner surface roughness on temperature profile in the well are predicted. This is in light of the predictions of the dependence of temperature on pressure and vice-versa.

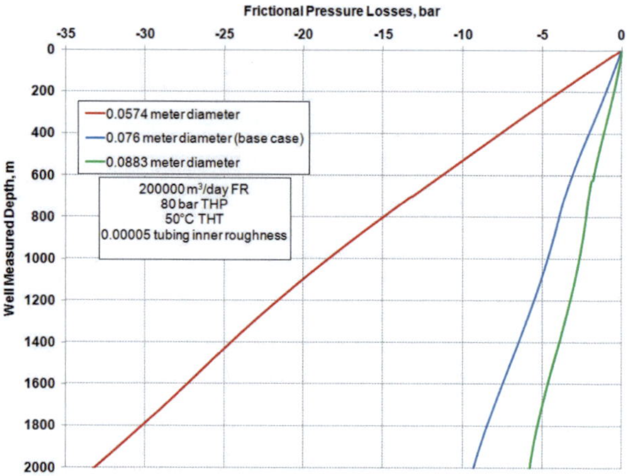

Figure 7.42: *Sensitivity of frictional pressure losses to tubing inner diameter*

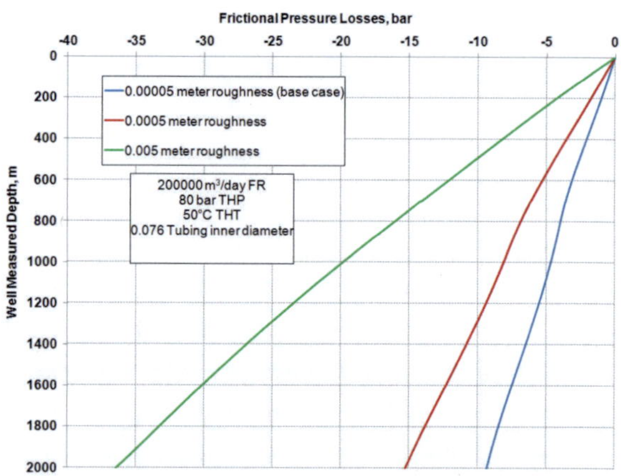

Figure 7.43: *Sensitivity of frictional pressure losses to tubing inner roughness*

Figure 7.44 and Figure 7.45 show the predictions of temperature profiles in the well as a function of tubing inner diameter and tubing inner surface roughness respectively. It appears that similar to the predictions of pressure profile; smaller diameters yield smaller temperatures in the well whereas smaller inner surface roughness results in higher temperatures. This corroborates the earlier provided explanations on the dependence of temperature on pressure through the Joule-Thomson effect and the heat of compression in the well. Higher pressures will thus induce higher temperatures.

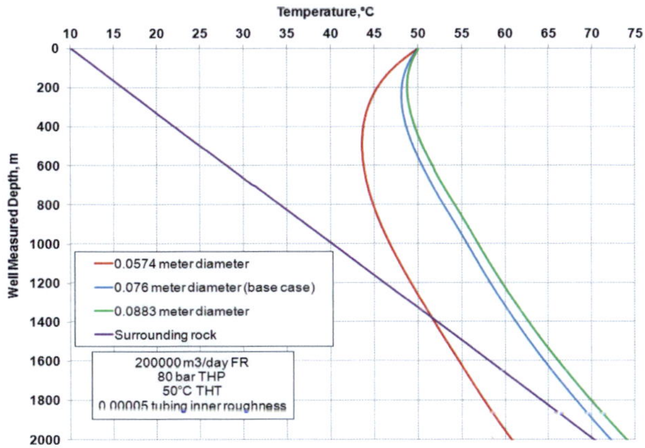

Figure 7.44: *Sensitivity of temperature to tubing inner diameter*

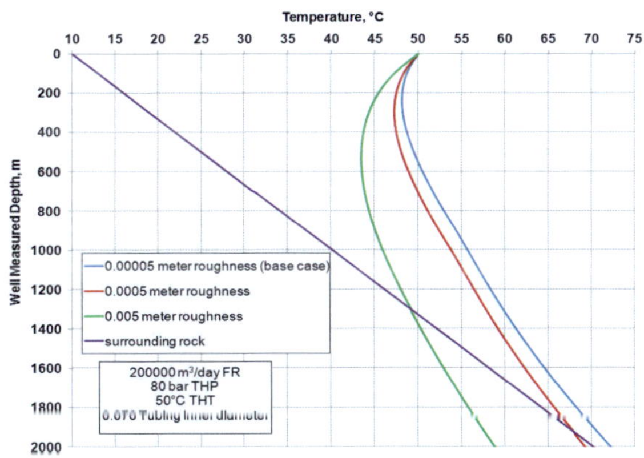

Figure 7.45: *Sensitivity of temperature to tubing inner roughness*

In conclusion it can be said that this work predicts that pressure and temperature in the well are smaller for smaller tubing inner diameter and larger for smaller tubing inner surface roughness. These are in agreement with the theory of flow into pipes and thermodynamics behavior of CO_2 implied during the development of the models earlier of this work.

7.4.5 Sensitivity of Pressure and Temperature to Injection Time

As injection proceeds and whenever the well is not thermally insulated, the fluid in the wellbore will exchange heat with the surrounding rock. This heat exchange process will result in either cooling or heating of the surrounding rock depending on whether it is hotter or cooler than the fluid at a particular depth. The heating or cooling effect will result in temperature changes which are more important at the rock-cement interface and propagate radially inside the rock until the rock temperature remains almost constant. This radial distance corresponds to the radius of thermal influence.. The sensitivity on the injection duration aims at analyzing how the model predicts the evolution of temperature and pressure of the fluid in the well as a function of injection days. It also enables to predict how the temperature propagates in the surrounding rock as a function of injection days. Emphasis is laid on the rock-cement interface, which is the junction point between the steady-state heat transfer process through the wellbore components (cement rings, casing, annulus, and tubing) up to the fluid and the transient heat transfer process through the rock up to the radius of thermal influence. Predictions are made over 90 days and reports are produced for 1, 5 and 10 days to reflect early time periods and 30, 60 and 90 days to reflect late time periods. The wellbore configuration is the same as the one presented in table 7.10. Two injection scenarios are considered. These are presented in table 15. They represent conditions similar to those of Ketzin (scenario I) and Snøhvit (scenario II) in order to allow investigations with different wellhead temperatures and allow for analysis of injection of gas and liquid phases separately. The rock thermal conductivity has been considered as 1.5 W/mK (base case) and annulus thermal conductivity 0.5 W/mK. The results are analyzed in the next section. Some difference with the Ketzin and Snøhvit results are however expected because of different thermal conductivities considered for rock and annulus. This is justified by the fact that the aim here is to investigate evolution of fluid temperature as well as propagation of temperature in the rock under different flow conditions, and to analyze sensitivities of these temperatures to rock thermal conductivity rather than to duplicate Ketzin or Snøhvit injection conditions.

***Table 7.15**: Summary of the base case scenarios used in investigating the sensitivity to tubing inner roughness*

Scenarios	Flow Rate	Temperature			Geothermal Gradient	Wellhead Pressure
		Wellhead	Reservoir	Surface		
	m^3/day	°C			°C/m	bar
I	30000	30	84	10	~0.037	60
II	500000	4	84	4	~0.04	100

7.4.5.1 Results and Discussion

Figure 7.46 shows the evolution of pressure and temperature in the wellbore as a function of injection time for scenario I.

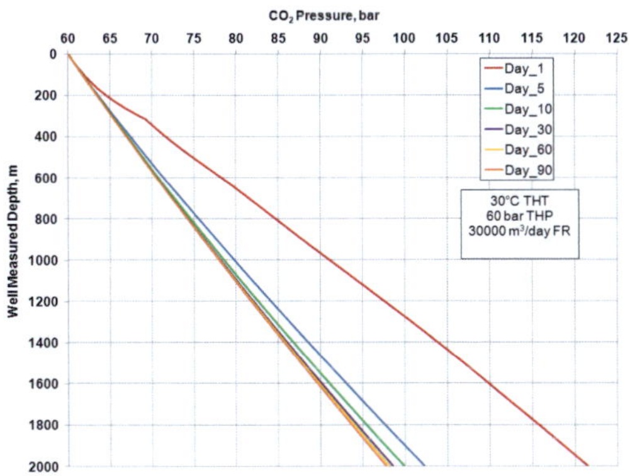

Figure 7.46: *Sensitivity of pressure to injection duration (scenario I)*

In this figure the pressure profile shows that considerably larger pressures are predicted for the first day. The bottomhole pressure in the first day is higher than that for the fifth day by approximately 20 bars. Thereafter from the fifth day, the difference reduces continuously up to the point that the predictions of the sixtieth and ninetieth day are almost similar. This could be explained by analyzing the temperature profiles in Figure 7.47.

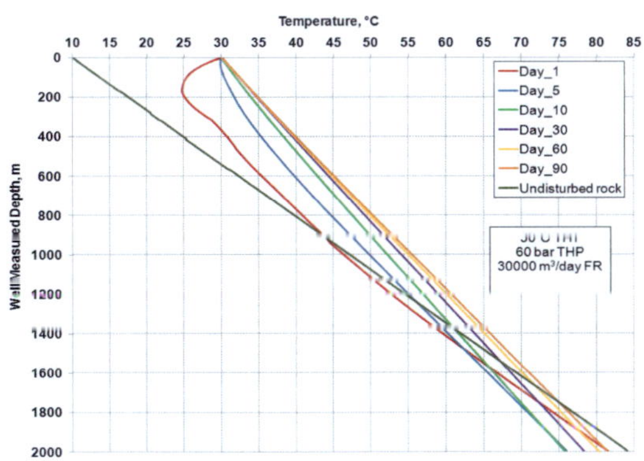

Figure 7.47: *Sensitivity of temperature to injection duration (scenario I)*

On the first day due to the relatively low flow rate, the fluid is significantly cooled by the surrounding rock in the upper part of the well causing predicted temperatures to be smaller compared with the days after. Fluid temperature reduces from 30°C at the wellhead to 24°C at around 150 meters. From this point the fluid is on the saturation line. As such the density is evaluated as the average of vapour and liquid densities, resulting in higher densities compared to those predicted for the days after. The fluid enters the supercritical region at around 400 meters in a phase much closer to liquid-like supercritical for this day than for the subsequent days. As a consequence the densities are larger and therefore the pressures predicted by the model are also larger. In the subsequent days the surrounding rock and the wellbore components have been heated by the fluid in the upper part of the well and the rock and wellbore components in the lower part of the well has been cooled by the fluid. The heating in the upper parts and the cooling in the lower parts continue up to the time where the fluid temperature becomes very close to the rock-cement interface temperature (see Figure 7.48). From this time the fluid and the rock are almost in thermal equilibrium and as such no significant amount of heat could be exchanged between them again.

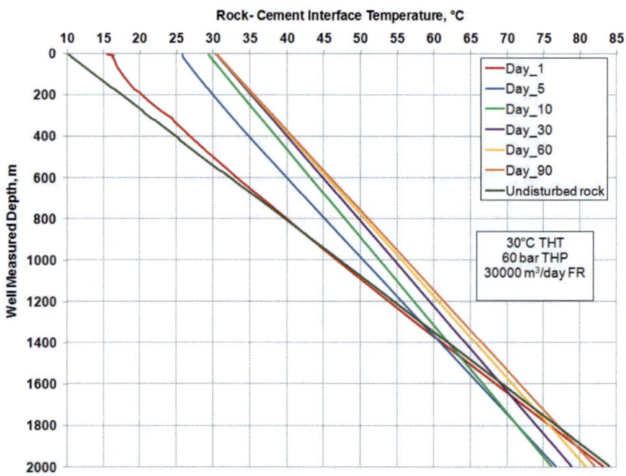

Figure 7.48: *Sensitivity of rock-cement interface temperature to injection duration (scenario I)*

This point is corroborated by the evolution of temperature in casing, tubing and annulus with time shown in Figure 7.49, Figure 7.50 and Figure 7.51 respectively. From the thirtieth day rock-cement interface, casing, tubing and annulus are all predicted to have almost the same temperature profile. As a consequence the difference in temperatures amongst them which is the driver to heat transfer approaches zero. It then justifies the thermal equilibrium that develops with time.

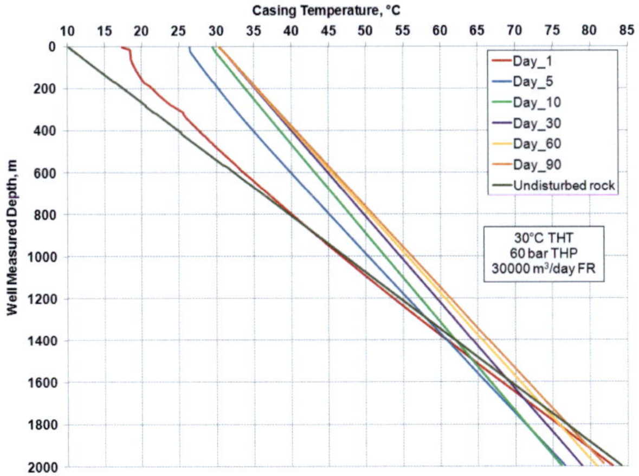

Figure 7.49: *Sensitivity of casing temperature to injection duration (scenario I)*

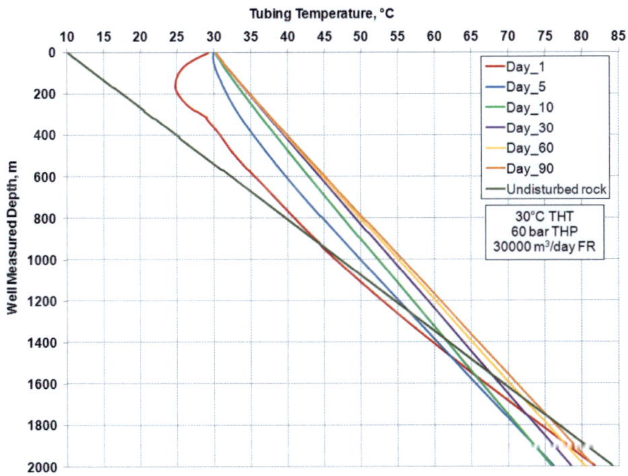

Figure 7.50: *Sensitivity of tubing temperature to injection duration (scenario I)*

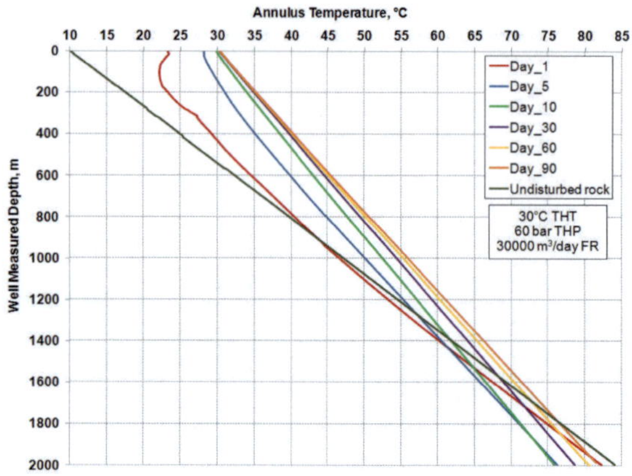

Figure 7.51: *Sensitivity of annulus temperature to injection duration (scenario I)*

The predictions of the pressure profiles for scenario II are shown in Figure 7.52. The difference in pressure predictions is not as pronounce as for scenario I. However smaller pressures are still predicted for the first day and the difference in predictions also diminishes gradually like with scenario I.

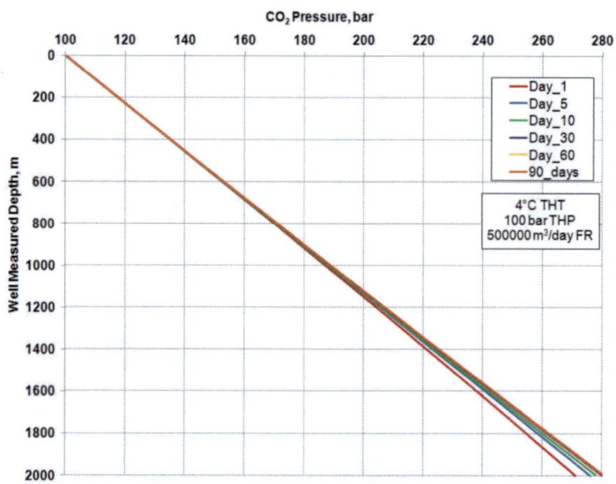

Figure 7.52: *Sensitivity of pressure to injection duration (scenario II)*

The temperature predictions can be seen in Figure 7.53. In the first day the rock heats the fluid. As a result its temperature decreases. The decrease in rock temperature continues and eventually thermal equilibrium is approached after the thirtieth day. As a result the pressure and temperature profiles for the sixtieth and the ninetieth days are almost similar. However in general unlike in scenario I where fluid pressure decreases with injection time in scenario II it increases with time. It is justified by the fact that the fluid temperature decreases with time in scenario II whereas it increases with time in scenario I.

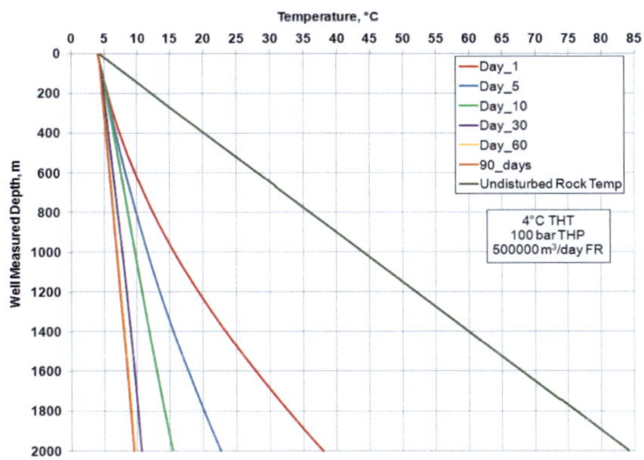

Figure 7.53: *Sensitivity of temperature to injection duration (scenario II)*

The fact that the thermal equilibrium is approached with injection time is illustrated by comparing the temperature of the fluid from the thirtieth day onward with the rock-cement interface temperature (showed in Figure 7.54) in same period. They are much closer to each other compared with other days.

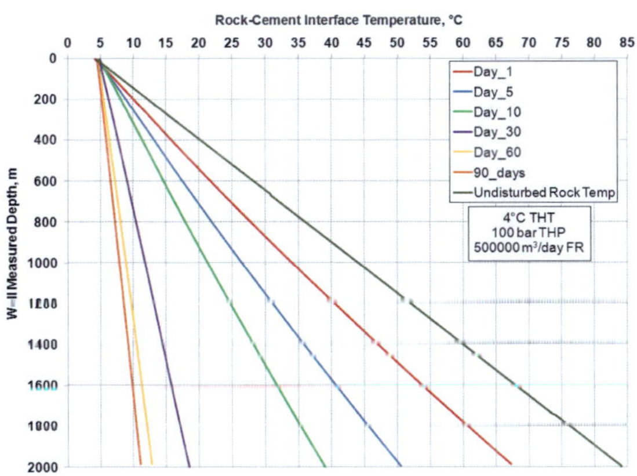

Figure 7.54: *Sensitivity of rock-cement temperature to injection duration (scenario II)*

The closeness of temperatures and the thermal equilibrium would be further confirmed by taking a look at the predictions of the profiles of casing, tubing and annulus temperature presented in Figure 7.55, Figure 7.56 and Figure 7.57 respectively. The casing temperature follows almost similar profile to rock-cement interface temperature whereas the tubing temperature is closer to the fluid temperature. As time goes on the difference amongst casing, annulus and tubing temperature decreases significantly to the point that no heat exchange would still be possible. Meanwhile the tubing temperature is approaching the fluid temperature and therefore also eliminates heat transfer between fluid and tubing.

In conclusion as injection goes on under constant injection rate and wellhead temperature and pressure, the heat exchanged between the surrounding rock and the injected CO_2 causes the temperature of the rock to approach that of the fluid in well. Eventually thermal equilibrium is approached and heat exchange becomes insignificant. The temperature of the system (fluid, wellbore components and rock) is then mostly governed by the heating effects due to compression of the fluid in the well.

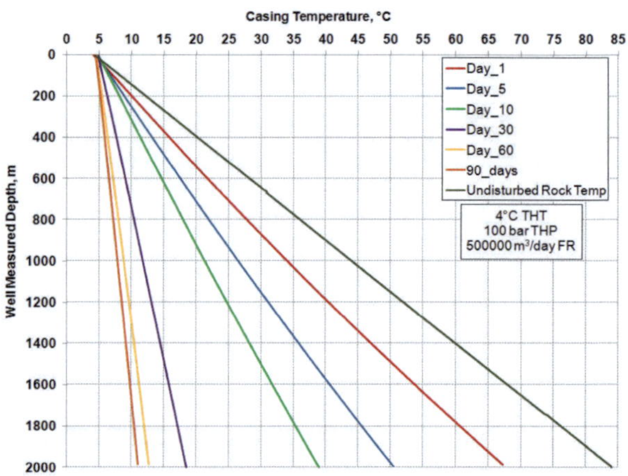

***Figure 7.55**: Sensitivity of casing temperature to injection duration (scenario II)*

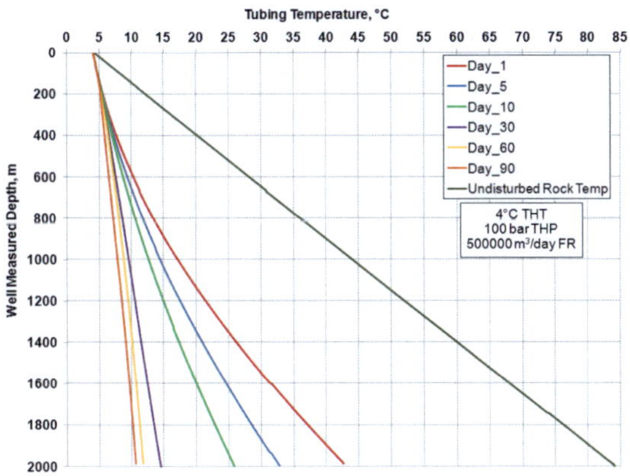

Figure 7.56: *Sensitivity of tubing temperature to injection duration (scenario II)*

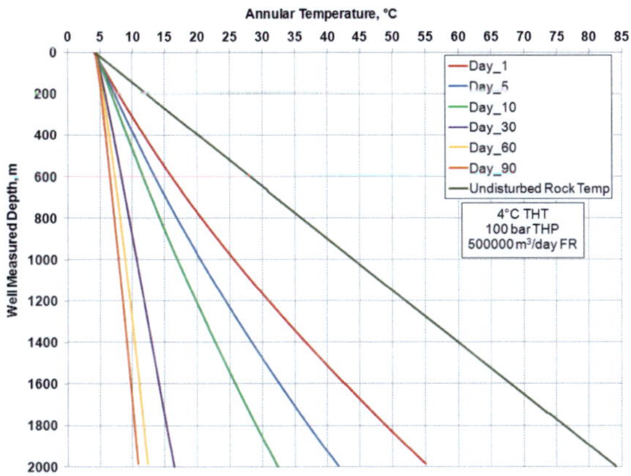

Figure 7.57: *Sensitivity of annulus temperature to injection duration (scenario II)*

7.4.6 Sensitivity of Radius of Thermal Influence and Surrounding Rock Temperature to Injection Duration

To check how the model predicts the radius of thermal influence (RTI) and analyze how the temperature evolves within this radius with time, ninety (90) days of injection are simulated. The foregone injection scenario I (with injection conditions close to those of Ketzin) and scenario II (with injection conditions close to those of Snoehvit) are also considered. The results are discussed in the following paragraphs.

7.4.6.1 Results and Discussion

Figure 7.58 shows the propagation of temperature within and the extent of the radius of thermal influence (RTI) with time for the injection scenario I at around 5 meters depth. At this depth the fluid is hotter than the surrounding rock and as such the rock is heated and its temperature increases with injection time. The increase in temperature is more pronounced at the interface with wellbore cement.

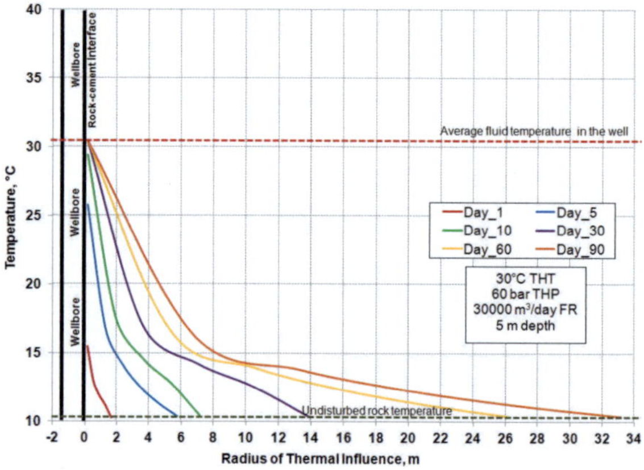

***Figure 7.58**: Sensitivity of rock temperature and radius of thermal influence to injection duration at the top of the wellbore (scenario I)*

With time the temperature of the rock-cement interface eventually gets very close to the fluid temperature. In this case this occurs from the thirtieth day and corroborates the attainment of thermal equilibrium explained in the preceding sections. As injection proceeds the RTI also increases. However the increase in temperature diminishes faster inward the surrounding rock until the RTI, where the rock temperature is unchanged. After ninety days of injection the RTI is predicted to be around 33 meters only. This can also be seen on Figure 7.59 and Figure 7.60 at 995 meters and 1995 meters respectively. At 995 meters the fluid temperature is also predicted to be higher than the surrounding rock temperature. As such the temperature propagation follows the same trends as at 5 meters. Contrarily at 1995 meters, the surrounding rock is hotter than the fluid. Therefore the rock is cooled so that its temperature approaches fluid temperature. Thermal equilibrium is also reached after the thirtieth day.

It is worthy to note that the average fluid temperature in these figures represents the average for the six days of which temperatures have been reported. It is also important to recall that the propagation of temperature is a function of thermal conductivity of the rock. The effect of rock thermal conductivity on temperature and RTI will been examined in next chapter dealing on the sensitivity of RTI and temperature propagation to the rock thermal conductivity.

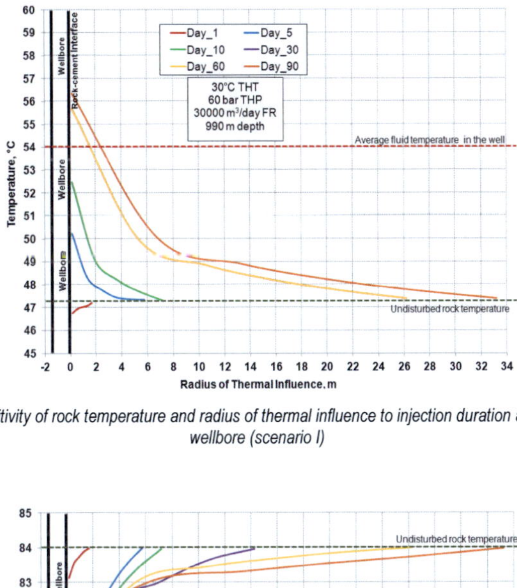

***Figure 7.59**: Sensitivity of rock temperature and radius of thermal influence to injection duration at the middle of the wellbore (scenario I)*

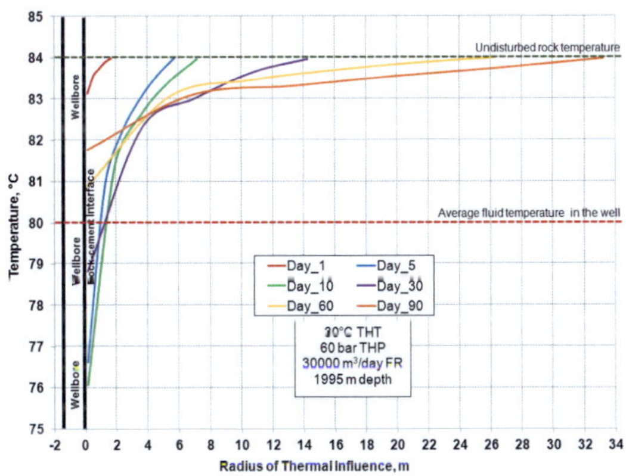

***Figure 7.60**: Sensitivity of rock temperature and radius of thermal influence to injection duration at the bottom of the wellbore (scenario I)*

In the case of scenario II the fluid temperature is just a little bit colder than the surrounding rock temperature at 5 meters depth. As such it can be seen in Figure 7.61 that the temperature propagation affects a radius of only up to a few meters in the rock and thermal equilibrium is attained already on the fifth day of injection.

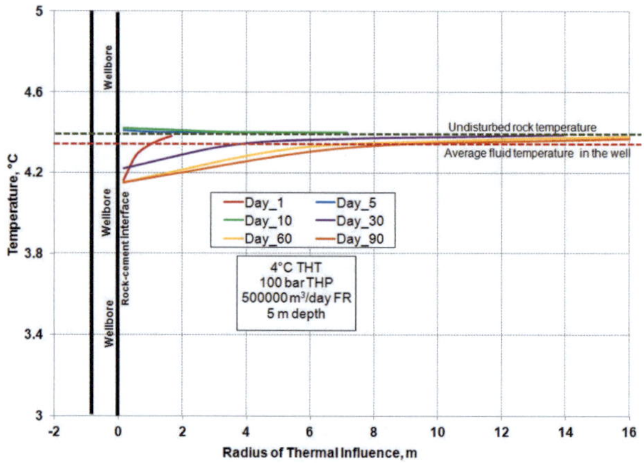

Figure 7.61: *Sensitivity of rock temperature and radius of thermal influence to injection duration at the top of the wellbore (scenario II)*

The faster attainment of thermal equilibrium and the smaller extent of RTI in this case would imply that the RTI as well as the attainment of thermal equilibrium are proportional to the difference between the fluid temperature at a particular depth and the geothermal temperature at that depth. This fact is corroborated with the extent of the RTI and the time of attainment of thermal equilibrium in Figure 7.62 and

Figure **7.63**. These figures represent the temperature propagation in the surrounding rock at 1000 meters and 1995 meters depths respectively. In this figure it can be seen the extent of RTI is predicted to be around 33 meters after 90 days of injection. This is similar to scenario I. The difference between fluid and geothermal temperature is around 30°C on the average whereas in scenario I it is around 8°C on the average. This would imply that below a certain difference between geothermal temperature and fluid temperature at a certain depth, the temperature difference stops having an effect the RTI or the attainment of thermal equilibrium. The estimation of this critical temperature difference will not be performed in this work because the qualitative conclusions are more important in this context where the focus is on the agreement of the codes with the physical principles.

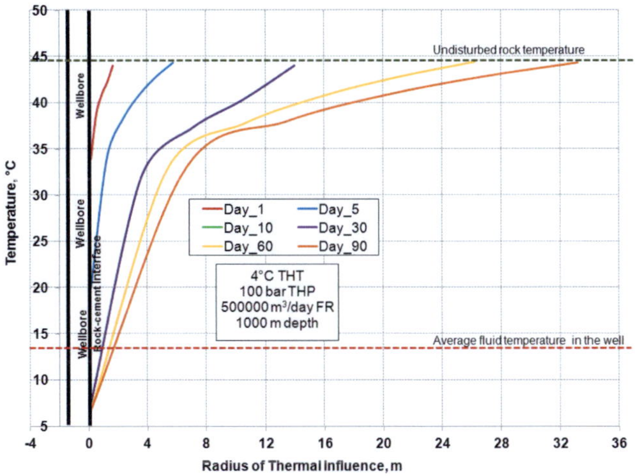

***Figure 7.62**: Sensitivity of rock temperature and radius of thermal influence to injection duration at the middle of the wellbore (scenario II)*

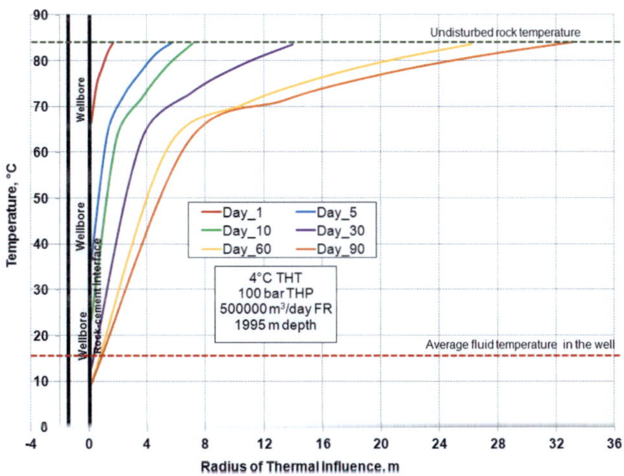

***Figure 7.63**: Sensitivity of rock temperature and radius of thermal influence to injection duration at the bottom of the wellbore (scenario II)*

At this point it would be worthy to look at how the model predicts temperature propagation in the rock for times below one week (early time) and after (later time). As discussed in chapter five, the time function for temperature propagation in the rock has been developed under the assumption that the well is an infinite cylindrical source of heat. As such the model would be able to make accurate predictions in early and late time, overcoming the early time limitations of Ramey's model which considers the well as a line source in an infinite reservoir. In Figure 7.59 to Figure 7.63 the predictions of rock temperature are reasonable for day 1 and day 5 and follow the trends of other days. They are also in agreement with the heat transfer theory. When the fluid is colder the rock temperature reduces as a result of loosing heat to the fluid and when the fluid is hotter the rock temperature increases as a result of taking up heat from the fluid. The trend is also the same as that for the later days. These would combine to further prove that the limitations of the Ramey's model have been overcome and the model make reasonable predictions form the first day of injection.

7.4.7 Sensitivity to Rock Thermal Conductivity

The sensitivity to rock thermal conductivity aims at analyzing how the model predicts the dependence of fluid temperature as well as the temperature propagation in the surrounding rock as a function of the thermal conductivity of the rock. To this end simulations were carried out with three different values of rock thermal conductivities: 1.5 W/mK (base case), 3.5 W/mK and 5.5 W/mK. The wellbore configuration and other parameters are the same as those in the scenario I and Scenario II of the preceding section. The results are reported for the first, tenth and thirtieth day respectively. They are discussed in the next paragraphs.

7.4.7.1 *Results and Discussion on Sensitivity of CO_2 Temperature and Pressure to Rock Thermal Conductivity*

Figure 7.64 shows the predictions of the temperature profiles in the well at different rock thermal conductivities for scenario I. On this figure it can be seen that, on the top of well, up to the depth of around 400 meters, the temperature of the fluid is by far higher than the initial temperature of the surrounding rock (geothermal temperature). Heat transfer between the fluid and the surrounding rock is very effective within this depth. The cases with 5.5 W/mK and 3.5 W/mK thermal conductivities predict faster heat transfer which result in faster cooling on the first day. This corresponds to lower temperatures predicted for these cases than for the base case with thermal conductivity of 1.5 W/mK.

Thereafter continuous injection causes the fluid and the surrounding rock to approach thermal equilibrium and have very close temperatures. The contrary behavior can be seen in the lower part of the well. In this part the fluid temperature becomes higher than the surrounding rock temperature after the first days of injection. As a consequence the rock is heated by the fluid. The temperature profiles show higher temperatures for the cases with 5.5 W/mK and 3.5 W/mK thermal conductivities. This implies that the higher the rock thermal conductivity the smaller the temperature difference that is needed to transfer the same amount of heat from wellbore to within the rock. This is corroborated by the evolution of the rock-cement interface temperature as a function of rock thermal conductivity shown on Figure 7.65. The reason is that the driver of temperature propagation in the rock is the difference between the rock-cement interface temperature and the undisturbed geothermal temperature. In this light, smaller temperatures are predicted for cases with higher rock thermal conductivities at the rock-cement interface in the upper part of the well where the fluid essentially heats the surrounding rock. In the same light, higher rock-cement interface temperatures are predicted for the cases with higher rock thermal conductivity in the lower part of the well. This confirms the fact that with higher thermal conductivity there is lower resistance to heat transfer within the rock and then smaller temperature drawdown is required for the transfer of the same rate of heat in or out of the surrounding rock.

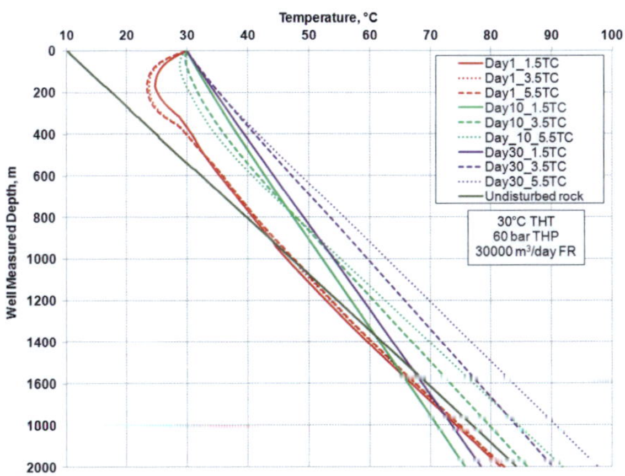

***Figure 7.64**: Prediction of sensitivity of CO2 temperature to rock thermal conductivity (scenario I)*

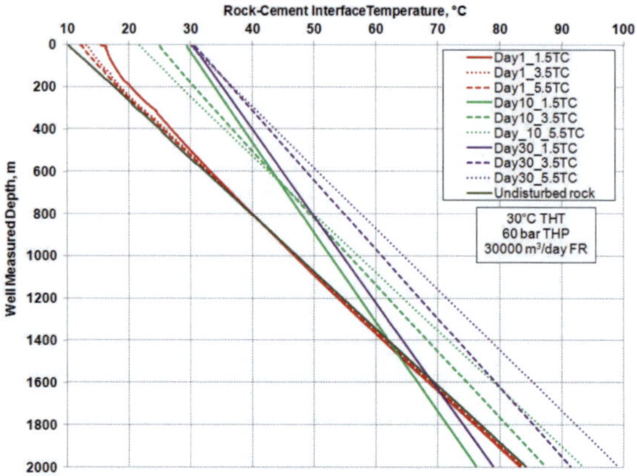

Figure 7.65: *Prediction of sensitivity of rock-cement interface temperature to rock thermal conductivity (scenario I)*

The effect of heat resistance would be better appreciated by looking at Figure 7.66 and

Figure 7.67 which show temperature propagation in the surrounding rock as a function of thermal conductivity at the top and bottom of the well respectively. In general the surrounding rock temperature is predicted to be higher for the cases with smaller thermal conductivities when the fluid heats the surrounding rock as in

Figure 7.67. This reflects the higher resistance to heat transfer which causes more accumulation of heat in the rock on the early days of injection before thermal equilibrium is approached. Once thermal equilibrium is approached like on the thirtieth day, the effect of thermal conductivity is no longer felt because there is barely no heat transferred in or out of the surrounding rock. In Figure 7.67 showing the same results for the bottom of the well, because the rock-cement interface temperature remains higher for cases with higher thermal conductivity, the temperature increases in the fluid very fast because smaller amount of heat will be extracted from it (see Figure 7.64). Note that the fluid in this scenario (scenario I) flows essentially in the gas phase. As such, it heats up significantly as a consequence of pressure increasing as the fluid descends down the well. The high temperature in the fluid and consequently at the rock-cement interface makes it that higher temperature are maintained in the rock as the system progresses towards thermal equilibrium. As a consequence the rock temperature remains always higher than that of the fluid for the case 1.5 W/mK thermal conductivity whereas the fluid temperature becomes higher that the rock temperature with time as injection proceeds for the cases with higher rock thermal conductivities. This can be confirmed by looking at Figure 7.64.

It is worthy to note that in all the cases the RTI is almost the same. This is explained by the fact that the RTI depends more on the amount of heat transferred to or from the fluid than on other parameters. It reflects the fact that in all the cases the same amount of heat is transferred to and from the fluid and only the induced temperature drawdown will change.

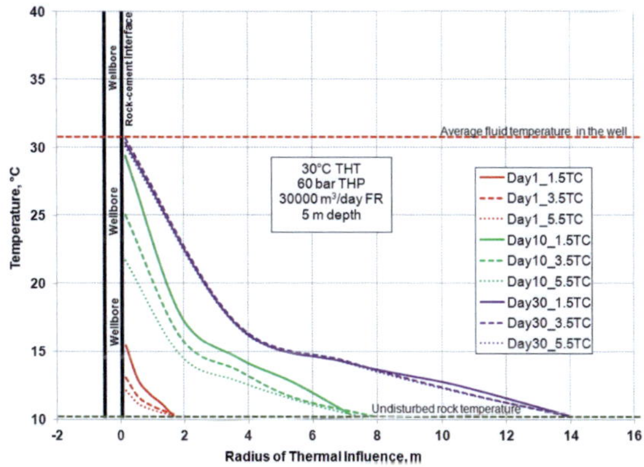

***Figure 7.66**: Prediction of sensitivity of temperature propagation in the rock to rock thermal conductivity at the top of the wellbore (scenario I)*

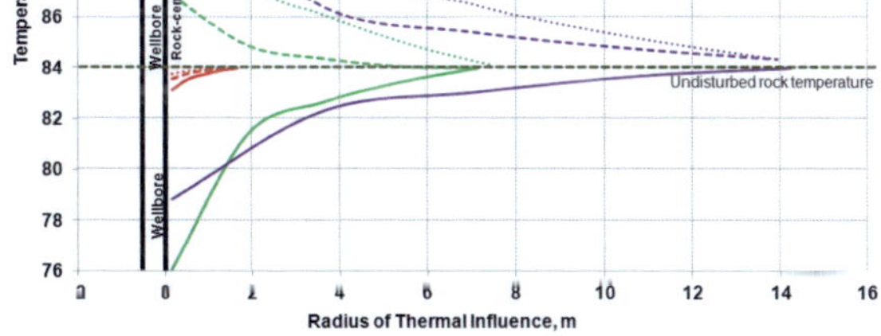

***Figure 7.67**: Prediction of sensitivity of temperature propagation in the rock to rock thermal conductivity at the bottom of the wellbore (scenario I)*

The predictions of pressure as a function of thermal conductivity for scenario I is presented in Figure 7.68. The pressure profiles reflect the temperature profile in Figure 7.64. On the first day where cases with higher rock thermal conductivities show lower temperature at the top of the well as a consequence of more intensive cooling, the pressures predicted are also higher in the well as a consequence of higher densities resulting from the lower temperatures that induced phase change. However as injection goes on the difference in densities diminishes reflecting no phase change or presence of two phases in the wellbore. As such the density becomes less sensitive to temperature changes. Therefore the pressure differences amongst the prediction become less pronounce. Nevertheless the tendency is that the cases with higher thermal conductivities will predict smaller pressures as a consequence of higher temperature predictions along the well.

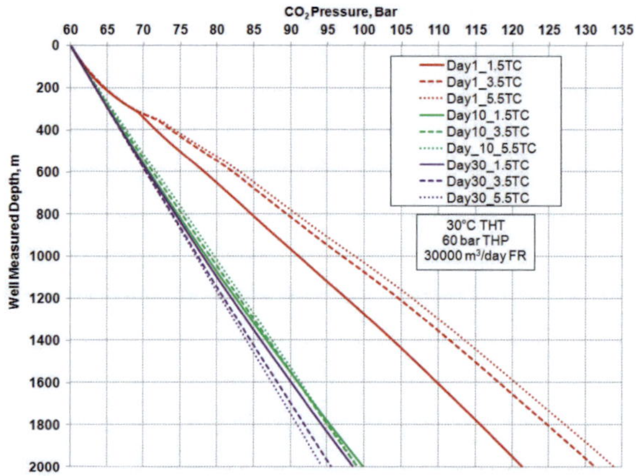

***Figure 7.68**: Prediction of sensitivity of CO2 pressure to rock thermal conductivity (scenario I)*

Before drawing plausible conclusions on the foregone points it would be important to examine the predictions for scenario II where the fluid temperature is by far smaller than that of the surrounding rock in the entire length of the well. Another point worthy to consider with scenario II is that the fluid flows essentially as liquid and liquid-like very dense supercritical fluid. As such the effect of heating due to compression is not as important as in scenario I.

Figure 7.69 shows the prediction of the temperature profiles as a function of the rock thermal conductivities for injection scenario II. Like with scenario I higher rock thermal conductivities cause lower drawdown in the surrounding and higher temperatures at the rock-cement interface. As such the fluid temperatures are higher for the cases with higher rock thermal conductivities in the present scenario, where the surrounding rock essentially heats up the fluid.

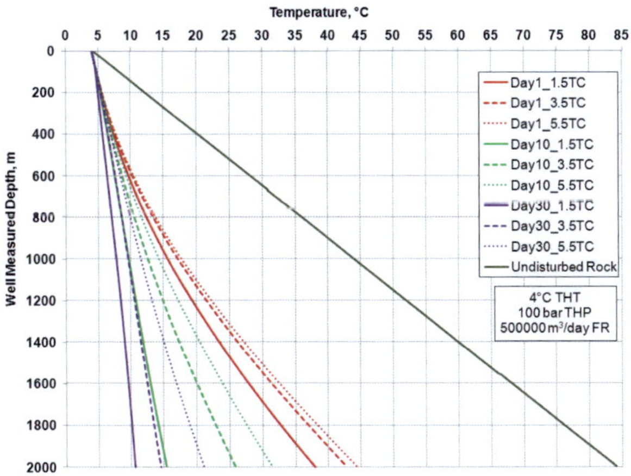

Figure 7.69: *Prediction of sensitivity of CO2 temperature to rock thermal conductivity (scenario II)*

The fact that smaller temperature drawdown for the cases with higher thermal conductivities will result in higher rock-cement interface temperature is presented in Figure 7.70. Because the same amount of heat is still transferred to the fluid (for all rock thermal conductivity cases), the end effect is the prediction of higher rock-cement interface temperatures and consequently that of higher fluid temperatures.

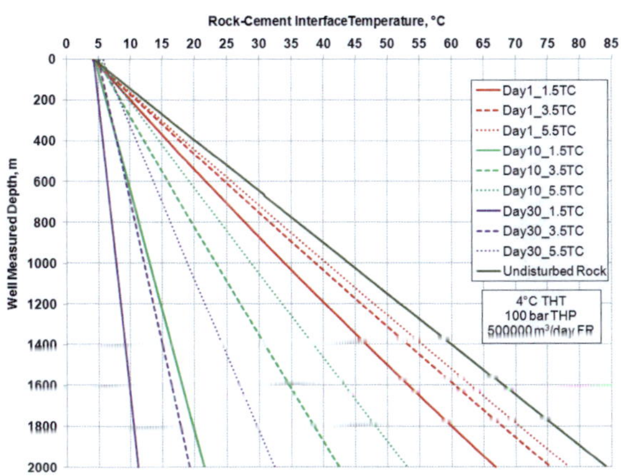

Figure 7.70: *Prediction of sensitivity of rock-cement interface temperature to rock thermal conductivity (scenario II)*

The effect of lower drawdown has also been discussed in making the same analysis for injection scenario I. It can further be corroborated by looking at the predictions of the temperature propagation in the rock as a function of thermal conductivities. These are presented in Figure 7.71 and Figure 7.72 for the middle and the bottom of the wellbore respectively

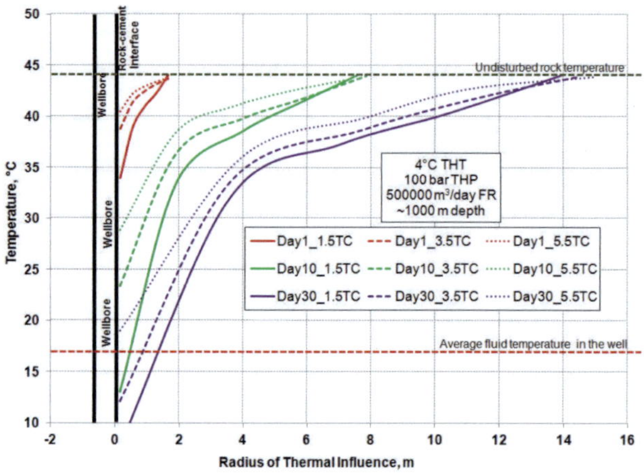

***Figure 7.71**: Prediction of sensitivity of temperature propagation in the rock to rock thermal conductivity at the middle of the wellbore (scenario II)*

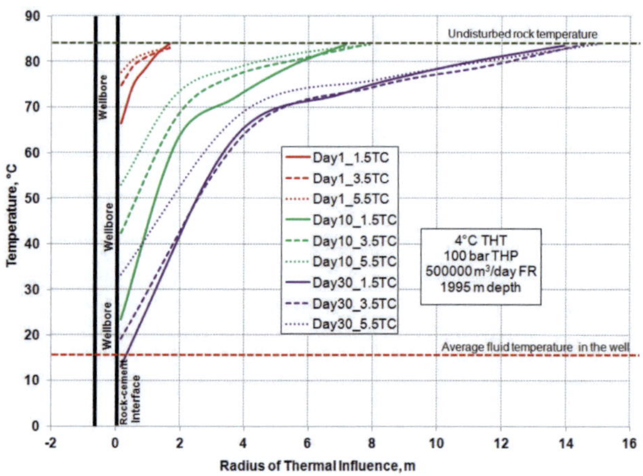

***Figure 7.72:** Prediction of sensitivity of temperature propagation in the rock to rock thermal conductivity at the bottom of the wellbore (scenario II)*

In analyzing the curves in these figures the general trend that appears is that in the present situation where the surrounding rock essentially heats up the fluid, higher rock thermal conductivities will yield lower decrease in temperature within the RTI. This is in fact another way to infer the earlier mentioned points on lower temperature drawdown corresponding to higher thermal conductivities as a result of smaller resistance to heat transfer within the rock.

The sensitivity of fluid pressure profile to rock thermal conductivity for scenario II is shown in Figure 7.73.

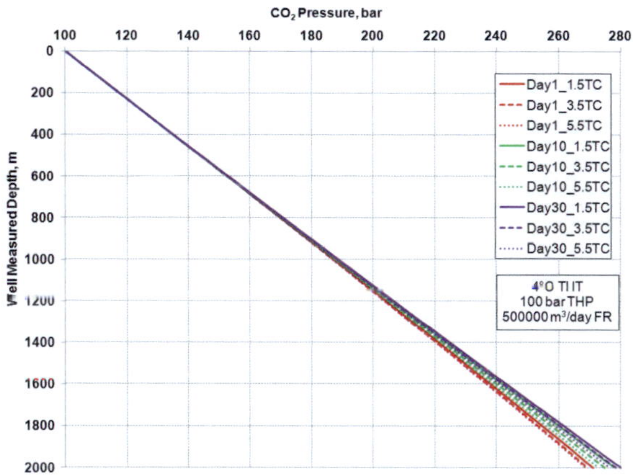

***Figure 7.73**: Prediction of sensitivity of CO_2 pressure to rock thermal conductivity (scenario II)*

Higher pressures are predicted for cases with smaller thermal conductivities. This corresponds to the predictions of smaller temperatures for these cases. However unlike in scenario I, the difference in pressure is not very high and occurs only after around 1200 meters. From this point the fluid is in the supercritical phase where densities are somehow smaller than liquid phase densities existing in the upper part of the well and as such sensitivity to temperature in the lower part becomes more important although not high. As a consequence higher temperatures would mean lower densities and therefore higher frictional pressure losses, especially for this scenario with a relatively high flow rate (500000 m^3/day). In this light the difference in bottomhole pressures caused by the difference in temperature would not only be attributed to different hydrostatic heads as a result of different densities but also to increasing frictional pressure losses with temperature.

8. CONCLUSIONS AND RECOMMENDATIONS

The effective injection of gases in geological formations such as saline aquifers, oil reservoirs and gas reservoirs require accurate prediction of the temperature and pressure along the well. This requirement becomes extremely meaningful for CO_2, which has critical temperature at 31.04°C and therefore can experience phase transition during injection. This would lead to abrupt changes in density and complex heat transport phenomena. As a consequence, the predictions of the bottomhole pressure and temperature would become difficult to make. Another important issue that renders predictions of pressure and temperature difficult is that as injection goes on with time, the heat transferred between the fluid and the surrounding rock changes the thermal equilibrium that initially existed between the well and the rock. As such it significantly alters the rate at which and eventually also the direction in which heat is transferred between the surrounding rock and the fluid with time. The foregone points justify the need for a robust procedure that will not only accurately estimate temperature and pressure along the well as a function of depth and time, but will also predict phase change and account for its effects on temperature and pressure along the well.

This work addresses this need with innovative approaches to analytical solutions of temperature and pressure equations on the one hand and on the other hand with the implementation of multiparameters equations of state which dramatically improve the accuracy of the thermodynamic and transport properties. These result in basic building blocks for rigorous prediction of temperature and pressure in an injection well for any gas in general. However because of the limitations of the equations of state, the models have been implemented in this context for CO_2 only.

Four temperature models have been developed. The main model is more rigorous in estimating the temperature along the well. It is non-adiabatic and as such it takes into account the temperature effects caused by compression/expansion of the fluid in the well and those related to the heat exchanged between the fluid and the surrounding rock. In this way the wellbore and the rock are assumed to form a heat exchanger with composite wall between the heat source and the heat sink. The composite wall is made up of the wellbore components (cement, casing, annulus and tubing).The three other models are the adiabatic model, the geothermal model and the isothermal model. The adiabatic model assumes that the well is perfectly insulated from the surrounding. Consequently only the effect of compression and/or expansion influences the temperature of the fluid in well. The geothermal model assumes that the temperature of the fluid is equal to that of the surrounding rock at each depth. This situation is therefore approached whenever the flow rate is low enough to allow longer contact time between the fluid and tubing surface. The isothermal model assumes that the temperature is uniform along the well.

It is approached in a horizontal well when inlet temperature is equal to geothermal temperature and frictional pressures losses are sufficiently small not to cause significant temperature drop.

In arriving at the solutions, the major thermodynamic and heat transfer processes that influence compressible fluids' flow in wellbores in general and CO_2 flow in particular have been considered. A greater emphasis has been laid on the accuracy of the estimation of the thermodynamic and transport properties of CO_2.

The accuracy with which these properties are estimated, especially the density, the viscosity and the thermal conductivity, strongly affects the predicted temperature and pressure. These properties are read from tables generated with fundamental multiparameters equations of state. These equations of state are generally used as benchmark to estimates made with commercial software in the absence of measured data. The Span and Wagner equation of state (Span and Wagner, 1996), is used for the calculation of density and other thermodynamic properties, whereas the Scalabrin et al's equation (Scalabrin et al., 2006) and the Fenghour et al 's equation (Fenghour et al., 1998) are used for the estimation of thermal conductivity and viscosity respectively. The tables cover the temperature range between 250 K and 450 K and pressure from atmospheric to 400 bar. The use of these tables enables to meet the dual and sometimes competing objectives of ensuring rapid calculations while preserving good accuracy. In fact these equations while being very accurate are very complex to code and require longer computation time than conventional cubic equations of state would take.

Beside the thermodynamic and transport properties, the temperature and pressure along the well also requires appropriate models and solutions that will minimize the errors in their predictions. The Bernoulli's equation modified to account for frictional pressure losses is implemented for pressure calculation. The enthalpy balance equation is implemented for temperature. The two equations are solved simultaneously and analytically to obtain fluid pressure and temperature along the well. The solution is based on dimensionless numbers. These dimensionless numbers govern hydrostatic pressure and frictional pressure losses in the well as well as the rate of heat transferred between the flowing fluid and the surrounding rock. As such they provide good control on the solution and the behavior of the model. They also ease the analysis of results and make it easy to identify user-related modeling errors.

The temperatures and the fluid properties are influenced by the heating in the wellbore. This is in turn in the main temperature model governed by two main processes: the compression/expansion of the fluid in the well and the exchange of heat between the fluid and the formation through the wellbore components (tubing, annulus, casing, and cement). To ensure that the heat propagation in the surrounding rock and therefore the temperature distribution in the fluid and the surrounding rock are rigorously calculated both for injection times smaller than one week and later, the analytical procedure proposed by Eppelbaum and Kutasov (2006) for temperature drawdown well testing has been adapted to be used in injection situations. The flowing fluid is assumed to be an infinite cylindrical heat source or heat sink. This enabled to derive a dimensionless time function valid for all times and as such to rigorously predict the radial propagation of the temperature in the rock surrounding the wellbore at early and late injection times.

Computer codes for the implementation of the models have been written on Microsoft Visual Studio platform using the Visual Basic.Net programming language. The codes can accommodate wellbore with varying deviation angles and varying wellbore diameters and tubing diameters. They can also accommodate layered rock formations with different thermal conductivities and thermal diffusivities. They are encapsulated in a user-friendly graphical user interface that eases the input and output of data to the user.

The computer code reads the fluid thermodynamic and transport properties at prevailing temperature and pressure in the well. When appropriate, it interpolates between tables' entries to estimate properties at intermediate points. The interpolations deter a little bit the accuracy of the estimates. However the deviations remain below 5% for all the phases of CO_2. As such the values obtained by interpolation are in general still better than estimates obtained with conventional equations of state, especially in the liquid and near critical regions.

Marching algorithm is implemented to minimize the errors related to the use of average values of thermodynamic and transport properties in finding analytical solutions of pressure and temperature in the well. This involves the division of the wellbore into segments of which lengths are determined by some validation criterion that ensures high accuracy of the solution. This criterion requires that the relative difference between the z-factor at the inlet of a segment and that at its outlet should not exceed 0.1%. A consequence of the segmentation of the well is the establishment of a computational loop which is repeated from the inlet to the outlet of the pipe. In this loop the fluid temperature, pressure and properties are assumed to be constant along each segment of the well. The temperature is calculated based on the values of pressure and fluid properties at the inlet of the segment. Thereafter the fluid properties are recalculated with the new temperature and the inlet pressure. Finally, the new fluid properties and temperature are used in calculating pressure at the outlet of the segment.

The analytical solutions to the models and the use of efficient marching algorithm combine to result in codes that reproduce field measured values with very high accuracy. The predictions of the code have been compared with measured data from two injection projects in saline aquifers under different injection conditions. The first project is Ketzin project, an onshore project in the north of Germany. CO_2 has been injected through a vertical well in an aquifer located at a depth of around 630 meters. In wellhead condition the CO_2 is in the gas phase and the injection takes place at approximately 30000 m^3/day. The fluid flows through the well mostly in the gas and gas like supercritical phases. The second project is Snøhvit project in Norway. In this project CO_2 has been injected though a deviated well offshore. The fluid is liquid at the wellhead and it is injected at a rate of around 500000 m^3/day. The flow through the well takes place in liquid and liquid like supercritical phase down the well to the reservoir located at a depth of around 2333 meters TVD from the sea floor. The well is slanted by approximately 8° in its upper part and 20° in its lower part.

In these two cases the relative deviations of simulated pressure values from the measured values are between 0.01% and 1.93 % while for the temperature values they are between 0.33% and 10.83%.

The rigorous prediction of temperature propagation has been used to estimate the thermal conductivity of the surrounding rock based on the simulated temperatures of CO_2 in the well. The surrounding rock thermal conductivities have been the main parameters used to match simulated values to measured ones. The best match for the Ketzin well has been obtained by considering that the surrounding rock has three layers with different thermal conductivities, 3.0 W/mK, 2.5 W/mK and 1.75 W/mK respectively. These thermal conductivities values are within the range reported for a reservoir in the Altensalzwedel area which is in the North German Basin like Ketzin. The best match was obtained for Snøhvit by assuming one rock layer with same thermal conductivity equal to 1.25 W/mK. The low thermal conductivity in one layer is in line with the fact that Snøhvit well is below water and may be surrounded by rock of fairly consolidated sand with high porosity and filled with water.

Further check on the models' accuracy has been conducted in comparing the predictions with the measured data from a CO_2 production well. The well has a long section slanted at 28.5°. The results are also very accurate. The maximum deviations of simulated values from measured ones are 3% and 4% for pressure and temperature respectively. It is however worthy to note that a general implementation of the code on production wells would require accounting for slip between phases in case of high liquid content. Phase slip has been neglected in this work.

Comparison of the predictions of the non-adiabatic model with those of the other temperature models provided confirmation of the full agreement of the models of this work with theory of fluid flow into cylindrical pipes and heat transfer in a heat exchanger with composite wall.. The results show that the predictions of the non-adiabatic model are close to those of the adiabatic model whenever the heat transferred between the fluid and the surrounding rock is negligible. They also reveal that the temperature predicted by the non-adiabatic model will be asymptotic to geothermal model at very low flow rates. Finally the non-adiabatic model predicts isothermal temperature in a horizontal well whenever inlet temperature is equal to surrounding rock temperature and the temperature drop due to frictional pressure losses is negligible. These are in agreement with heat transfer process in a heat exchanger with composite wall discussed by Coulson and Richardson (1996) and with the theory fluid flow in wells presented by Brill and Mukherjee (1999) .

Further confirmation of the agreement with above-mentioned theories has been obtained through the analysis of the sensitivities of the predictions of the non-adiabatic temperature model to injection conditions (flow rate and wellhead temperature and pressure.), injection time and tubing inside diameter and inside surface roughness.

The higher the flow rate, the higher the frictional pressure losses and therefore the lower the pressures along the well and the lower the effects of compression on fluid temperature. Higher flow rates also result in lower rate of heat exchange between the fluid and the surrounding rock. The effect of wellhead temperature on temperature and consequently pressure in the well depends also on flow rate and the difference between the wellhead temperature and surrounding rock temperature on the top of the well. For higher flow rates higher wellhead temperatures will yield higher fluid temperatures along the well. This is a consequence of smaller influence of the surrounding rock. Therefore risks of phase transition would also be low. The higher the wellhead pressure the higher the pressure along the well. However the difference in pressures in the well is not the same as the difference in wellhead pressures. This is because the higher the density the smaller the frictional losses and all other parameters being equal, the density change is very sensitive the pressure change and vice-versa. As such higher wellhead pressures also resulted in higher hydrostatic heads.

Sensitivities to tubing inside diameter suggested that the larger the diameter the smaller the frictional losses and hence the higher the predicted pressure and temperature. The sensitivity to tubing inside surface roughness showed contrary effect. The model predicted higher frictional pressure losses for higher roughness.

As injection proceeds it is predicted that the surrounding rock temperature and the fluid temperature will get closer to each other to the point that heat transferred between fluid and rock will become negligible. Thermal equilibrium is reached at this point. The model will behave like the adiabatic model and consequently fluid temperature will be controlled only by the compression or expansion of the fluid in the well. This is in agreement with the Fourier's law of heat transfer by conduction in solids and the Newton's law of cooling for which the driving force for heat transfer is the temperature difference between the media involved in the process.

To complete the sensitivity analysis, further sensitivities of the temperatures of the fluid and the temperature propagation in rock to thermal conductivities of the rock have been performed. It is predicted that the value of rock thermal conductivity, the time at which thermal equilibrium would be reached. The value of rock thermal conductivity has almost no influence on the radius of thermal influence. It is predicted that the RTI depends more on the difference between fluid temperature and rock temperature at a certain depth (hence rate of heat transferred) and on the injection time. Again in agreement with the temperature drawdown well testing theory discussed by Eppelbaum and Kutasov (2006).

While highlighting the accuracy of the present work in predicting temperature and pressure of CO_2 in an injection well, it is also important to recognize the room for improvements in many of its aspects. These are closely related to the assumptions underpinning the models.

One of these assumptions is very conservative and considers that the vertical heat propagation in the surrounding rock and in the fluid is to be neglected. Another one that could have some sensitive implications on the predictions is that the heat of vaporization and the heat liberated upon condensation in case of phase transition are assumed to propagate all into the formation through the wellbore components and over the entire length of the well. This is not totally true because a part of this heat will be carried over in the descending the fluid by convection. Furthermore the mass exchange between phases whenever the gas phase encounters the liquid phase in the well is not taken into consideration. Consequently the models will not be able to simulate temperature and pressure profiles in as shut-in well experiencing boiling and condensation with acceptable accuracy. This is of importance because shut-in situations will always occur during real field injection.

In the light of the foregoing it is recommended that, further work would consist of modeling the mass transfer and heat transfer phenomena with analogy to the phenomena at play in a distillation column. It would also be important to take into consideration the vertical propagation of the temperature in the rock. Another further work, which could also significantly improve the accuracy of these models would consist of directly implementing the multi-parameter equations of state in the codes by taking advantage of the high speed of today's and future's processors to develop time-efficient codes with these equations. The codes would also be generalized for other gases or mixtures of gases by including tables of properties generated with dedicated multiparameters equations of state or by direct implementation of these equations of state. Including phase slip through for instance drift-flux models will enable to also predict temperature and pressure in production wells producing gases with high liquid content.

In the light of the foregoing the final conclusion is that although this work has room for improvements, it has shown to perform very well in predicting temperature and pressure of CO_2 in an injection well as function of depth and time. It has also shown to accurately predict temperature and pressure in a CO_2 production with low liquid content, where phase slip is negligible.

9. REFERENCES

AL-SYABI, Z., DANESH, A., TOHIDI, B., TODD, A. C. & TEHRANI, D. H. 2001. A residual viscosity correlation for predicting the viscosity of petroleum reservoir fluids over wide ranges of pressure and temperature. *Chemical Engineering Science,* 56, 6997-7006.

ALVES, I. N., ALHANATI, F. J. S. & SHOHAM, O. 1992. A Unified Model for Predicting Flowing Temperature Distribution. in Wellbores and Pipelines. *SPE Production Engineering,* 7, 363-367.

ANIS, M. & BUTHOD, P. 1974. How Flashing Fluid Change Phase in Pipeline. *Oil and Gas Journal,* 150.

ANSARI, A. M., SYLVESTER, N. D., SARICA, C., SHOHAM, O. & BRILL, J. P. 1994. A Comprehensive Mechanistic Model for Upward Two-Phase Flow in Wellbores. *SPE Production & Operations,* 9, 143-151.

ATKINS, P. W. 1986. *Physical Chemistry,* Oxford.

AYALA, O. F., AYALA, L. F. & AYALA, O. M. 2007. Multi-phase Flow Analysis in Oil and Gas Engineering Systems and its Modeling. *Hydrocarbon World* 2007, 57-61.

BAHADORI, A. 2008. New correlation accurately predicts thermal conductivity of liquid paraffin hydrocarbons. *Journal of the Energy Institute,* 81, 59-61.

BAHADORI, A. & VUTHALURU, H. 2009. A Simple Method for Prediction of Transport Properties of Carbon Dioxide. *Asia Pacific Health, Safety, Security and Environment Conference.* Jakarta, Indonesia: Society of Petroleum Engineers.

BAHONAR, M., AZAIEZ, J. & CHEN, Z. J. 2010. Two Issues in Wellbore Heat Flow Modeling along with the Prediction of Casing Temperature in the Steam Injection Wells. *Abu*

Dhabi International Petroleum Exhibition and Conference. Abu Dhabi, UAE: Society of Petroleum Engineers.

BEGGS, H. D. & BRILL, J. P. 1973. A Study of two-phase Flow in Inclined Pipes. *Journal of Petroleum Technology,* May 1973, 607-617.

BOYLE, B., CARROLL, J. J. 2001. Calculation of Acid Gas Density in the Vapor, Liquid and Dense Region-Phase Regions. Available: http://www.gasliquids.com/papers/agidensity.pdf.

BRILL, J. P. & MUKHERJEE, H. 1999. *Multphase Flow in Wells,* Richardson, TX, SPE.

CHERIC. 2010. *Korea Thermophysical Properties Data Bank* [Online]. Seoul: CHERIC. Available: http://www.cheric.org/research/kdb/hcprop/showprop.php?cmpid=1943 [Accessed 22th November 2010].

CHRISTIE, B. 2007. Analysis of the Frictional Heating Effect In Pipelines. Pipeline Simulation Interest Group.

CHUNG, T. H., AJLAN, M., LEE, L. L. & STARLING, K. E. 1988. Generalized multiparameter correlation for nonpolar and polar fluid transport properties. *Industrial & Engineering Chemistry Research,* 27, 671-679.

CISMONDI, M. & MOLLERUP, J. 2005. Development and application of a three-parameter RK-PR equation of state. *Fluid Phase Equilibria,* 232, 74-89.

CIVAN, F. 2008. A correlation for thermal conductivity of liquid n-alkanes based on the Vogel-Tammann-Fulcher-Hesse equation. *Chemical Engineering Science,* 63, 5883-5886.

COULSON, J. M. & RICHARDSON, J. F. 1996. *Coulson & Richardson's chemical engineering,* Oxford; Boston, Butterworth-Heinemann.

COULTER, D. M. & BARDON, M. F. 1979. revised equation improved flowing gas temperature prediction *Oil and Gas Journal,* (Feb. 26, 1979), 107-108.

CRONSHAW, M. B. & BOLLING, J. D. 1982. A Numerical Model of the Non-Isothermal Flow of Carbon Dioxide in Wellbores. *SPE California Regional Meeting.* San Francisco, California: 1982 Copyright 1982, Society of Petroleum Engineers of AIME.

DBI 2011. Contributions to the CLEAN Project's Final Report. Freiberg.

DDBST. 2012. *VAP(Vapor Pressure) Data for Carbon Dioxide* [Online]. Dortmund. Available: http://ddbonline.ddbst.de/EE/1050%20VAP%20%28Vapor%20Pressure%29.shtml [Accessed 19th February 2012].

DUNS, H. A. R., N. C. J 1963. Vertical flow of gas and liquid mixtures in wells. *6th World Petroleum Congress.* Frankfurt am Main, Germany: World Petroleum Congress.

DURRANT, A. J. & THAMBYNAYAGAM, R. K. M. 1986. Wellbore Heat Transmission and Pressure Drop for Steam/Water Injection and Geothermal Production: A Simple Solution Technique. *SPE Reservoir Engineering,* 1, 148-162.

DUSCHEK, W., KLEINRAHM, R. & WAGNER, W. 1990a. Measurement and correlation of the (pressure, density, temperature) relation of carbon dioxide I. The homogeneous gas and liquid regions in the temperature range from 217 K to 340 K at pressures up to 9 MPa. *The Journal of Chemical Thermodynamics,* 22, 827-840.

DUSCHEK, W., KLEINRAHM, R. & WAGNER, W. 1990b. Measurement and correlation of the (pressure, density, temperature) relation of carbon dioxide II. Saturated-liquid and saturated-vapour densities and the vapour pressure along the entire coexistence curve. *The Journal of Chemical Thermodynamics,* 22, 841-864.

ECONOMIDES, M. J., HILL, D. A. & EHLIG-ECONOMIDES, C. 1993. *Petroleum Production Systems,* Upper Saddle River, Prentice Hall.

EDALAT, M., BOZAR-JOMEHRI, R. B. & MANSOORI, G. A. 1993. Generalized equation predicts vapor pressure of hydrocarbons. *Oil and Gas Journal,* 91:5, Pages: 39-40.

ELY, J. F. & HANLEY, H. J. M. 1983. Prediction of transport properties. 2. Thermal conductivity of pure fluids and mixtures. *Industrial & Engineering Chemistry Fundamentals,* 22, 90-97.

ELY, J. F., MAGEE, J. W. & HAYNES, W. M. 1987. Thermophysical properties for special high CO2 content mixtures. Tulsa: Gas Processors Association.

EPPELBAUM, L. V. & KUTASOV, I. M. 2006. Pressure and temperature drawdown well testing: similarities and differences. *Journal of Geophysics and Engineering,* 3, 12.

FAN, T.-B. & WANG, L.-S. 2006. A viscosity model based on Peng-Robinson equation of state for light hydrocarbon liquids and gases. *Fluid Phase Equilibria,* 247, 59-69.

FAROUQ ALI, S. M. 1981. A Comprehensive Wellbore Stream/Water Flow Model for Steam Injection and Geothermal Applications. 21, 527-534.

FENGHOUR, A., WAKEHAM, W. A. & VESOVIC, V. 1998. The Viscosity of Carbon Dioxide. *Journal of Physical and Chemical Reference Data,* 27, 31-44.

FONTANILLA, J. P. & AZIZ, K. 1982. Prediction of Bottom-hole Conditions For Wet Steam Injection Wells. 21.

FREDENSLUND, A., JONES, R. L. & PRAUSNITZ, J. M. 1975. Group-contribution estimation of activity coefficients in nonideal liquid mixtures. *AIChE Journal,* 21, 1086-1099.

FREY, K., AUGUSTINE, C., CICCOLINI, R. P., PAAP, S., MODELL, M. & TESTER, J. 2007. Volume translation in equations of state as a means of accurate property estimation. *Fluid Phase Equilibria,* 260, 316-325.

FURUKAWA, H., SHOHAM, O., BRILL, J. P. 1986. Predicting Compositional Two-Phase Flow behavior in Pipelines. *ASME Journal of Energy Resources and Technology,* 108.

GOULD, T. L. 1979. Compositional Two-Phase Flow in Pipelines. *SPE Journal of Petroleum Technology,* 31, 373-384.

GRADSHTEYN, I. S. & RYZHIK, I. M. 2007. *Table of Integrals, Series and Products,* New York.

GREGORY, G. & AZIZ, K. 1978. Calculation of Pressure And Temperature Profiles In Multiphase Pipelines And Simpne Pipeline Networks. 17.

GUO, X. Q., SUN, C. Y., RONG, S. X., CHEN, G. J. & GUO, T. M. 2001. Equation of state analog correlations for the viscosity and thermal conductivity of hydrocarbons and reservoir fluids. *Journal of Petroleum Science and Engineering,* 30, 15-27.

GUO, X. Q., WANG, L. S., RONG, S. X. & GUO, T. M. 1997. Viscosity model based on equations of state for hydrocarbon liquids and gases. *Fluid Phase Equilibria,* 139, 405-421.

HAGEDORN, A. R. & BROWN, K. E. 1965. Experimental Study of Pressure Gradients Occurring During Continuous Two-Phase Flow in Small-Diameter Vertical Conduits. *SPE Journal of Petroleum Technology,* 17, 475-484.

HAGOORT, J. 2004. Ramey's Wellbore Heat Transmission Revisited. *SPE Journal,* 9, 465-474.

HAGOORT, J. 2005. Prediction of wellbore temperatures in gas production wells. *Journal of Petroleum Science and Engineering,* 49, 22-36.

HASAN, A. R. & KABIR, C. S. 1991. Heat Transfer During Two-Phase Flow in Wellbores; Part I--Formation Temperature. *SPE Annual Technical Conference and Exhibition.* Dallas, Texas: 1991 Copyright 1991, Society of Petroleum Engineers, Inc.

HASAN, A. R. & KABIR, C. S. 1994. Aspects of wellbore heat transfer during two-phase flow. *Journal Name: SPE Production and Facilities (Society of Petroleum Engineers); (United States); Journal Volume: 9:3*, Medium: X; Size: Pages: 211-216.

HEIN, M. A. Year. Rigorous and Approximate Methods for CO2 Pipeline Analysis in Facilities, Pipelines and Measurements: A Workbook for Engineers. *In:* 41st Petroleum Mechanical Engineering Workshop and Conference, 1985 1985 Kansas City, MO, USA.

HENNINGES, J., LIEBSCHER, A., BANNACH, A., BRANDT, W., HURTER, S., KÖHKER, S., MÖLLER, F. & GROUP, C. S. 2010. P-T-rho and Two-Phase Fluid Conditions with Inverted Density Profile in Observation wells at the CO2 Storage Site at Ketzin (Germany). *Energy Procedia.*

HERNING, F. & ZIPPERER, L. 1936. Beitrag zur Berechnung der Zähigkeit Gasgemische aus den Zähigkeitwerten der Einzelbestandteile. *Das Gas und Wasserfach,* 5, 69-73.

HOLMES, J. A., BARKVE, T. & LUND, O. 1998. Application of a Multisegment Well Model to Simulate Flow in Advanced Wells. *European Petroleum Conference.* The Hague, Netherlands: 1998 Copyright 1998, Society of Petroleum Engineers Inc.

JOSSI, J. A., STIEL, L. I. & THODOS, G. 1962. The viscosity of pure substances in the dense gaseous and liquid phases. *AIChE Journal,* 8, 59-63.

KABACK, W., M. AND STARLING, K. E. Year. A Four-Parameter Corresponding States Correlation for Fluid Compressibility Factors. *In:* Oklahoma Academy of Science, 1976. 125-132.

KABIR, C. S. & HASAN, A. R. 1990. Performance of a two-phase gas/liquid flow model in vertical wells. *Journal of Petroleum Science and Engineering,* 4, 273-289.

KATZ, D. L. 1972. Thermodynamic Analysis of Frictional Flow Effects in Pipeline Flow. *Oil & Gas Journal* March 1972.

KIM, Y. 2007. Equation of State for Carbon Dioxide. *Journal of Mech. Sc. & Tech,* 21, 316-325.

KLOTZ, I. M. 1964. *Introduction to Chemical Thermodynamics,* New York.

KREYSZIG, E. 1999. *Advanced Engineering Mathematics,* New York, John Wiley & Sons.

LEE, B. I. & KESLER, M. G. 1975. A generalized thermodynamic correlation based on three-parameter corresponding states. *AIChE Journal,* 21, 510-527.

LINGEN, P. L. 1974. Description of Groningen Gas Well Performance Suitable for Medium and Long Term Planning. *SPE European Spring Meeting.* Amsterdam, Netherlands: 1974 Copyright 1974, American Institute of Mining, Metallurgical, and Petroleum Engineers, Inc.

LOHRENZ, J., BRAY, B. G. & CLARK, C. R. 1964. Calculating Viscosities of Reservoir Fluids From Their Compositions. *SPE Journal of Petroleum Technology,* 16, 1171-1176.

LOLLEY, C. S. & RICHARDSON, W. C. 1998. Minimum Pseudocomponent Requirements for Compositional Thermal Simulation of. *SPE/DOE Improved Oil Recovery Symposium.* Tulsa, Oklahoma.

LOULI, V., BOUKOUVALAS, C., VOUTSAS, E., MAGOULAS, K. & TASSIOS, D. 2007. Application of the UMR-PRU model to multicomponent systems: Prediction of the phase behavior of synthetic natural gas and oil systems. *Fluid Phase Equilibria,* 261, 351-358.

LU, M. & CONNELL, L. D. 2008. Non-isothermal flow of carbon dioxide in injection wells during geological storage. *International Journal of Greenhouse Gas Control,* 2**,** 248-258.

MATHIAS, P. M., NAHEIRI, T. & OH, E. M. 1989. A density correction for the Peng--Robinson equation of state. *Fluid Phase Equilibria,* 47**,** 77-87.

MATHIAS, P. M. & PAREKH, V. S. 1996. Prediction and Correlation of the Viscosity and Thermal Conductivity of Pure Fluids and Mixtures. *AICHE Annual Meeting.* IL, Chigago, USA.

MATHIAS, P. M., PAREKH, V. S. & MILLER, E. J. 2001. Prediction and Correlation of the Thermal Conductivity of Pure Fluids and Mixtures, Including the Critical Region. *Industrial & Engineering Chemistry Research,* 41**,** 989-999.

MCCOLLUM, D. L. & OGDEN, J. M. 2006. Techno-Economic Models for Carbon Dioxide Compression, Transport and Storage and Correlations for Estimating Carbon Dioxide Density and Viscosity. Davis: Institute of Transport Studies-University of California.

MCDONALD, C. M. & FLOUDAS, C. A. 1995. Global Optimization and Analysis for the Gibbs Free Energy Function Using the UNIFAC, Wilson, and ASOG Equations. *Industrial & Engineering Chemistry Research,* 34**,** 1674-1687.

MEHROTRA, A. K., MONNERY, W. D. & SVRCEK, W. Y. 1996. A review of practical calculation methods for the viscosity of liquid hydrocarbons and their mixtures. *Fluid Phase Equilibria,* 117**,** 344-355.

NITO DE CASTRO, C. A. & WAKEHAM, W. A. 1996. Methodology. *In:* MILLAT, J., DYMOND, J. H., NIETO DE CASTRO, C. A (ed.) *Transport Properties of Fluids: Their Correlation, Prediction and Estimation.* Cambridge: IUPAC.

NORDEN, B., FÖRSTER, A., BEHRENDS, K., KRAUSE, K., STECKEN, L. & MEYER, R. 2012. Geological 3-D model of the larger Altensalzwedel area, Germany, for temperature prognosis and reservoir simulation. *Environ Earth Science,* 62 (2012)**,** 511-526.

ORKISZEWSKI, J. 1967. Predicting Two-Phase Pressure Drops in Vertical Pipe. *SPE Journal of Petroleum Technology,* 19**,** 829-838.

ORR JR, F. M. 2007. *Theory of Gas Injection Processes*, Tie-Line Technology.

PADUA, A., WAKEHAM, W. A. & WILHELM, J. 1994. The viscosity of liquid carbon dioxide. *International Journal of Thermophysics,* 15**,** 767-777.

PAN, H. & FIROOZABADI, A. 1998. Complex Multiphase Equilibrium Calculations by Direct Minimization of Gibbs Free Energy by Use of Simulated Annealing. *SPE Reservoir Evaluation & Engineering,* 1**,** 36-42.

PÁTEK, J., KLOMFAR, J., CAPLA, L. & BURYAN, P. 2005. Thermal Conductivity of Carbon DioxideMethane Mixtures at Temperatures Between 300 and 425 K and at Pressures up to 12MPa. *International Journal of Thermophysics,* 26**,** 577-592.

PATEL, N. C. & TEJA, A. S. 1982. A new cubic equation of state for fluids and fluid mixtures. *Chemical Engineering Science,* 37**,** 463-473.

PATERSON, L., LU, M., CONNELL, L. & ENNIS-KING, J. P. 2008. Numerical Modeling of Pressure and Temperature Profiles Including Phase Transitions in Carbon Dioxide Wells. *SPE Annual Technical Conference and Exhibition.* Denver, Colorado, USA: Society of Petroleum Engineers.

PEDERSEN, K. S. & CHRISTENSEN, P. L. 2007. *Phase behavior of petroleum reservoir fluids,* Boca Raton, CRC/Taylor & Francis.

PENELOUX, A., RAUZY, E. & FREZE, R. 1982. A Consistent Correction for Redlich-Kwong-Soave Volumes. *Fluid Phase Equilib.*, 8(1982), 7-23.

PENG, D., ROBINSON, D. B. 1976. A New Two-Constant Equation of State. *Industrial and Engineering Chemistry,* Fundamentals 15, 59-64.

POURAFSHARY, P., VARAVEI, A., SEPEHRNOORI, K. & PODIO, A. 2008. A Compositional Wellbore/Reservoir Simulator to Model Multiphase Flow and Temperature Distribution. *International Petroleum Technology Conference.* Kuala Lumpur, Malaysia: International Petroleum Technology Conference.

PUSCH ET AL 2010. Common Features of Carbon Dioxide and Underground Gas Storage. *European Oil & Gas Magazine,* 36, 131-136.

RAMEY JR., H. J. 1962. Wellbore Heat Transmission. *SPE Journal of Petroleum Technology,* 14, 427-435.

REDLICH, O., KWONG, N. S 1949. On the Thermodynamics of Solutions. V: An Equation of Sate. Fugacities of Gaseous Solutions. *Chem. Rev.,* 44, 233-244.

REID, R. C., PRAUSNITZ, J. M. & POLING, B. E. 1987. *The properties of gases and liquids.*

ROBERTSON, E. C. 1988. Thermal Properties of Rocks. *Open-File Report 88-441 of the US Geological Survey.* Reston, Virginia.

SASTRI, S. R. S. & RAO, K. K. 1999. A new temperature-thermal conductivity relationship for predicting saturated liquid thermal conductivity. *Chemical Engineering Journal,* 74, 161-169.

SATTER, A. 1965. Heat Losses During Flow of Steam Down a Wellbore. *SPE Journal of Petroleum Technology,* 17, 845-851.

SCALABRIN, G., MARCHI, P., FINEZZO, F. & SPAN, R. 2006. A Reference Multiparameter Thermal Conductivity Equation for Carbon Dioxide with an Optimized Functional Form. *Journal of Physical and Chemical Reference Data,* 35, 1549-1575.

SETZMANN, U. & WAGNER, W. 1991. A New Equation of State and Tables of Thermodynamic Properties for Methane Covering the Range from the Melting Line to 625 K at Pressures up to 100 MPa. *Journal of Physical and Chemical Reference Data,* 20, 1061-1155.

SHENG, W., CHEN, G. J. & LU, H. C. 1989. Prediction of transport properties of dense gases and liquids by the Peng-Robinson (PR) equation of state. *International Journal of Thermophysics,* 10, 133-144.

SINGHE, A. T. & GANZER, L. 2010. On The Ability of Cubic Equations of State to Estimate CO2 Density. *DGMK-Tagungsbericht,* 1, 521-531.

SOAVE, G. 1972. Equilibrium constants from a modified Redlich-Kwong equation of state. *Chemical Engineering Science,* 27, 1197-1203.

SPAN, R. & WAGNER, W. 1996. A New Equation of State for Carbon Dioxide Covering the Fluid Region from the Triple-Point Temperature to 1100 K at Pressures up to 800 MPa. *Journal of Physical and Chemical Reference Data,* 25, 1509-1596.

STATOIL 2009. *MeMo: CO2 Injection from Melkoeya.*

SUZUKI, J. I. & UEMATSU, M. 2002. Equation of state for fluid mixtures based on the principle of corresponding states with a two-fluid model: Application to fluid mixtures of water + ammonia. *Heat Transfer—Asian Research,* 31, 320-330.

TWU, C. H., COON, J. E. & BLUCK, D. 1998. Comparison of the Peng–Robinson and Soave–Redlich–Kwong Equations of State Using a New Zero-Pressure-Based Mixing

Rule for the Prediction of High-Pressure and High-Temperature Phase Equilibria. *Industrial & Engineering Chemistry Research,* 37, 1580-1585.

TWU, C. H., SIM, D., TASSONE, V. 2002. Getting a Handle on Advanced Cubic Equations of State. *Chemical Engineering Progress*, 58-65.

ÜLKER, E. B. 2009. *Investigation of the CO2 Storage Capacity of Aquifer Structures: CO2 Storage in a Bundsanstein Prototype Aquifer.* Doctor of Enginering Dissertation.

VALDERRAMA, J. O. 2003. The State of the Cubic Equations of State. *Industrial & Engineering Chemistry Research,* 42, 1603-1618.

VESOVIC, V., ET AL 1990. The Transport Properties of carbon Dioxide. *J. Phys. Chem. Data,* 19, 763-908.

VOUTSAS, E., MAGOULAS, K. & TASSIOS, D. 2004. Universal Mixing Rule for Cubic Equations of State Applicable to Symmetric and Asymmetric Systems: Results with the Peng–Robinson Equation of State. *Industrial & Engineering Chemistry Research,* 43, 6238-6246.

WAGNER, W. 1973. New vapour pressure measurements for argon and nitrogen and a new method for establishing rational vapour pressure equations. *Cryogenics,* 13, 470-482.

WAKEHAM, W. A. & NIETO DE CASTRO, C. A. 1996. Technological Importance. *In:* MILLAT, J., DYMOND, J. H., NIETO DE CASTRO, C. A (ed.) *Transport Properties of Fluids: Their Correlation, Prediction and Estimation.* Cambridge: IUPAC.

WALAS, M. S. (ed.) 1985. *Phase equilibria in chemical engineering,* Boston: Butterworth.

WARK, K. & RICHARDS, D. E. 1999. *Thermodynamics,* Boston, McGraw-Hill.

WILLHITE, G. P. 1967. Over-all Heat Transfer Coefficients in Steam And Hot Water Injection Wells. *SPE Journal of Petroleum Technology,* 19, 607-615.

WOOLEY, G. R. 1980. Computing Downhole Temperatures in Circulation, Injection, and Production Wells. *SPE Journal of Petroleum Technology,* 32, 1509-1522.

WU, Y.-S. & PRUESS, K. 1990. An Analytical Solution for Wellbore Heat Transmission in Layered Formations (includes associated papers 23410 and 23411). *SPE Reservoir Engineering,* 5, 531-538.

ZÉBERG-MIKKELSEN, C. K., BAYLAUCQ, A., BARROUHOU, M., QUIÑONES-CISNEROS, S. E. & BONED, C. 2004. Comparative study of viscosity models on the ternary system methylcyclohexane + cis-decalin + 2,2,4,4,6,8,8-heptamethylnonane up to 100 MPa. *Fluid Phase Equilibria,* 222-223, 135-148.

APPENDIX A- Further Calculations

Section A.1: Fugacity Coefficient and K-Values Calculation

The calculation of K-values is based on the equality of fugacity between phases and follows the relationships that exist between Gibb's free energy, chemical potential and fugacity. It is inspired by the procedure proposed by Atkins (1986 , Pp. 127&128).

For pure substances, the chemical potential can be expressed with equation A.1.1 so long as the substance is a perfect gas or behaves almost like a perfect gas.

$$\lambda = \lambda_{std} + RT \ln\left(\frac{P}{P_{std}}\right) \quad \text{A.1.1}$$

Where,

λ is the chemical potential

λ_{std} is the chemical potential and standard pressure, which is 1 atm.

P is the prevailing pressure

P_{std} is the atmospheric pressure

R is the universal gas constant

T is the prevailing absolute temperature

In the case of a real gas, this equation takes the form of A.1.1 below, where f is the fugacity and represents an effective pressure of the system. Therefore it has the same unit as pressure.

$$\lambda = \lambda_{std} + RT \ln\left(\frac{f}{P_{std}}\right)$$

A.1.2

Where,

The Gibb's free energy G at a certain pressure P_2 can be found as a function of its value at another pressure P_1 with equation A.1.3 , provided that the system's volume V is known.

$$G(P_2) = G(P_1) + \int_{P_1}^{P_2} V dP \quad \text{A.1.3}$$

In the case of a perfect gas, the molar volume V is related to the pressure by equation A.1.4.

$$V = \frac{RT}{P} \quad \text{A.1.4}$$

Replacing the volume in equation A.1.3 by equation A.1.4 yield equation A.1.5.

$$G(P_2) = G(P_1) + RT\int_{P_1}^{P_2} \frac{dP}{P} \qquad \text{A.1.5}$$

Integrating equation A.1.5 and setting $P_1 = P_{std}$ to reflect standard condition, the Gibb's free energy of a perfect gas at pressure P can be obtained as a function of its value at standard pressure with equation A.1.6 below.

$$G(P) = G_{std} + RT\ln\left(\frac{P}{P_{std}}\right) \qquad \text{A.1.6}$$

Recalling that for a pure substance, the Gibb's free energy is equal to the chemical potential, which reflects the tendency of a system to shift in the direction of decreasing Gibb's free energy (Atkins, 1986 , P. 127), equation A.1.6 equivalent to A.1.1 in terms of chemical potential. Moreover, if account is taken for the deviation from ideality of a real gas, A.1.6 will turn to A.1.2, in terms of chemical potential.

From equation A.1.1 it appears that the chemical potential at any pressure is composed of its value at standard pressure and a residual term (second term on the right hand side) that accounts for pressure change. Because at standard pressure, all gases behave almost ideally, the shift from ideality of a real gas would be reflected by the residual term in the chemical potential equation. To account for the shift, the pressure can be replaced by fugacity to obtain equation A.1.7.

$$\int_{P_1}^{P_2} VdP = \lambda(P_2) - \lambda(P_1) = RT\ln\left(\frac{f_2}{f_1}\right) \qquad \text{A.1.7}$$

For a perfect gas, equation A.1.7 will become A.1.8.

$$\int_{P_1}^{P_2} VdP = \lambda_{IG}(P_2) - \lambda_{IG}(P_1) = RT\ln\left(\frac{P_2}{P_1}\right) \qquad \text{A.1.8}$$

Where, index IG denotes ideal gas.

Subtraction of equation A.1.8 from A.1.7 will yield equation A.1.9.

$$\ln\left(\frac{f_2/P_2}{f_1/P_1}\right) = \frac{1}{RT}\int_{P_1}^{P_2} (V - V_{IG})dP \qquad \text{A.1.9}$$

Assuming that the gas would behave ideally at standard pressure (1 atm.) where, fugacity and pressure are equal, setting $P_1 = P_{std}$ and removing the index 2 on P_2, equation A.1.9 will becomeA.1.10. Equation A.1.10 represents the expression of the fugacity coefficient, ϕ , which is the ratio of the fugacity of the system to its pressure.

$$\ln\phi = \frac{1}{RT}\int_{P_{std}}^{P} \left(V - V_{IG}\right) dP \qquad \text{A.1.10}$$

Where,

$$\phi = \frac{f}{P} \qquad \text{A.1.11}$$

Equation A.1.4 can be rearranged to give A.1.12.

$$PV_{IG} - RT = 0 \qquad \text{A.1.12}$$

For a real gas, A.1.12 will not equate to zero, for deviation from the ideal gas behavior is reflected on the volume. Let's represent this deviation with δ in equation A.1.13.

$$PV - RT = \delta \qquad \text{A.1.13}$$

Subtracting equation A.1.12 from A.1.13 and rearranging would yield:

$$PV - PV_{IG} = \delta \qquad \text{A.1.14}$$

Putting back equation A.1.4 into A.1.14 and dividing by pressure yields:

$$\Phi = V - \frac{RT}{P} \qquad \text{A.1.15}$$

Where,

$$\Phi = \frac{\delta}{P} = V - V_{IG} \qquad \text{A.1.16}$$

Putting equation A.1.16 into A.1.10 gives:

$$\ln\phi = \frac{1}{RT}\int_{P_{std}}^{P} \Phi dP \qquad \text{A.1.17}$$

Putting equation A.1.16 in A.1.17 gives:

$$\ln \phi = \frac{1}{RT}\int_{P_{std}}^{P}\left(V - \frac{RT}{P}\right)dP \qquad \text{A.1.18}$$

Integration of Equation A.1.18 yields:

$$\ln \phi = \frac{V}{RT}(P - P_{std}) - \ln\left(\frac{P}{P_{std}}\right) \qquad \text{A.1.19}$$

Therefore,

$$\phi = Exp\left[\frac{V}{RT}(P - P_{std}) - \ln\left(\frac{P}{P_{std}}\right)\right] \qquad \text{A.1.20}$$

Working in SI units and approximating P_{std} to 1 bar, equation A.1.21 becomes:

$$\phi = Exp\left[\frac{V}{RT}(P-1) - \ln P\right] \qquad \text{A.1.21}$$

In terms of liquid and vapor properties:

$$\phi_L = Exp\left[\frac{V_L}{RT}(P-1) - \ln P\right] \qquad \text{A.1.22}$$

$$\phi_v = Exp\left[\frac{V_v}{RT}(P-1) - \ln P\right] \qquad \text{A.1.23}$$

The K values can now be estimated with equation A.1.24

$$K = \frac{\phi_L}{\phi_V} = \frac{\chi}{x} \qquad \text{A.1.24}$$

Recalling compositional material balance equation,

$$\chi = 1 - x \qquad \text{A.1.25}$$

Therefore,

$$K = \frac{\chi}{1-\chi} \qquad \text{A.1.26}$$

Equation A.1.24 can be used to calculate K-values, from which the fraction of fluid in the gas phase and that in the liquid phase can be obtained, by rearranging equation A.1.26 to give equation A.1.27.

$$\chi = \frac{K}{K+1} \qquad \text{A.1.27}$$

$$x = 1 - \chi \qquad \text{A.1.28}$$

Section A.2: Calculation of Radius of Thermal Influence

For a borehole, assuming a cylindrical source, the wall temperature can be calculated by analogy with the bottomhole flowing pressure expression under similar assumption. Eppelbaum and Kutasov (2006) therefore proposed an equation to this end, which has been modified to suit the situation of a wellbore containing a flowing or stationary fluid that acts as a heat source or sink (equation A.2.1).

$$T_{fc} = T(t, r_w) = T_i + \frac{q}{2\pi k_R} \ln\left[1 + \left(c - \frac{1}{a + \sqrt{t_D}}\right)\sqrt{t_D}\right] \qquad \text{A.2.1}$$

Where,

$$t_D = \frac{k_R t}{c_{p_R} \rho_R r_w^2} = \frac{\alpha t}{r_w^2} \qquad \text{A.2.2}$$

$$a = 2.7010505 \qquad \text{A.2.3}$$

$$c = 1.4986055 \qquad \text{A.2.4}$$

$$\alpha = \frac{k_R}{c_{p_R} \rho_R} \qquad \text{A.2.5}$$

Rearranging equation A.2.1 as a function of heat flow per unit time yields equation A.2.6

$$q = \frac{2\pi k_R (T_{fc} - T_i)}{\ln\left[1 + \left(c - \frac{1}{a + \sqrt{t_D}}\right)\sqrt{t_D}\right]} \qquad \text{A.2.6}$$

The dimensionless cumulative heat flow per unit length in the wellbore can be calculated in analogy with the cumulative fluid production calculation in pressure well testing, with equation A.2.7.

$$Q_D = \int_1^{\infty} T_{Dr}\left(r_D, t_D\right) r_D dr_D \tag{A.2.7}$$

Where,

$$T_{Dr} = \frac{T(r,t) - T_i}{T_{fc} - T_i} \tag{A.2.8}$$

$$r_D = \frac{r}{r_w} \tag{A.2.9}$$

Equation A.2.7 can be re-written in terms dimensionless radius (r_D) and dimensionless radius of thermal influence (R_{in}) as in equation A.2.10.

$$Q_D = \int_1^{\infty} \left(1 - \frac{\ln r_D}{\ln R_{in}}\right) r_D dr_D \tag{A.2.10}$$

Under the constraints that: $R_{in} > r_D$.

Where,

$$R_{in} = \frac{r_{in}}{r_w} \tag{A.2.11}$$

Integration of equation A.2.10 using table of integrals of Gradshteyn and Ryshik (2007) will give equation A.2.12

$$Q_D = \left[\frac{1}{4}\frac{R_{in}^2 - 2\ln\left(R_{in}\right) - 1}{\ln\left(R_{in}\right)}\right] \tag{A.2.12}$$

The cumulative heat flow in the segment of length ΔL is given by equation A.2.13..

$$Q = qt = -2\pi\rho_R c_{p_R} r_w^2 \left(T_{fc} - T_i\right) Q_D \tag{A.2.13}$$

Combination of equations A.2.1, A.2.2, A.2.12 and A.2.13 will give equation A.2.14 that expresses the dimensionless time (t_D) as a function of the dimensionless radius of thermal influence (R_{in}).

$$t_D = \frac{1}{4}\frac{R_{in}^2 + 2\ln R_{in} - 1}{\ln R_{in}} \ln\left[1 + \left(c - \frac{1}{a + \sqrt{t_D}}\right)\sqrt{t_D}\right] \qquad \text{A.2.14}$$

Equation A.2.14 is implicit in t_D, Therefore it is solved iteratively with t_D estimated by equation A.2.2, changing the values of R_{in} until convergence is reached.

APPENDIX B –Thermodynamic and Transport Properties in the Tables

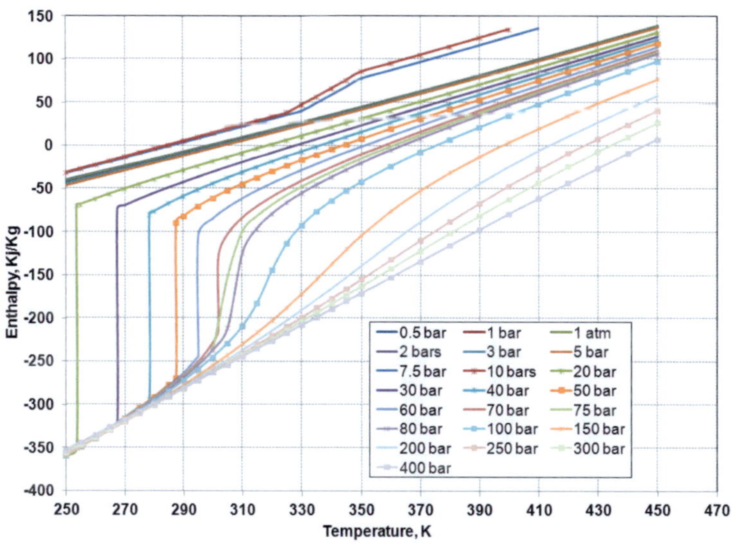

***Figure B.1**: Isobars of enthalpy vs. temperature in the thermodynamic properties' tables*

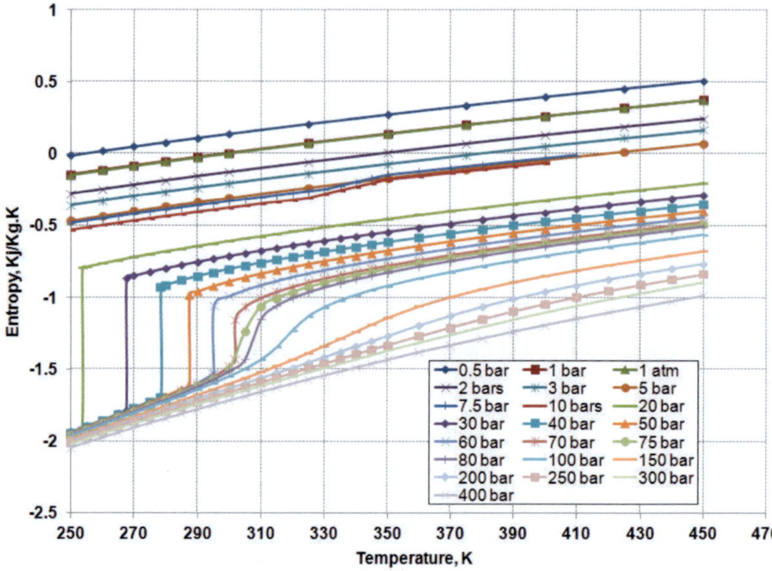

***Figure B.2**: Isobars of entropy vs. temperature in the thermodynamic properties' tables*

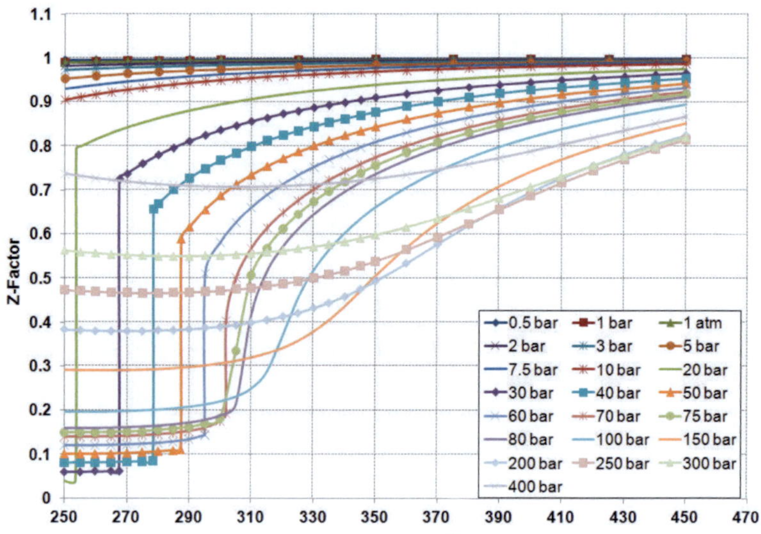

***Figure B.3**: Isobars of Z-factor Vs. temperature in the thermodynamic properties tables*

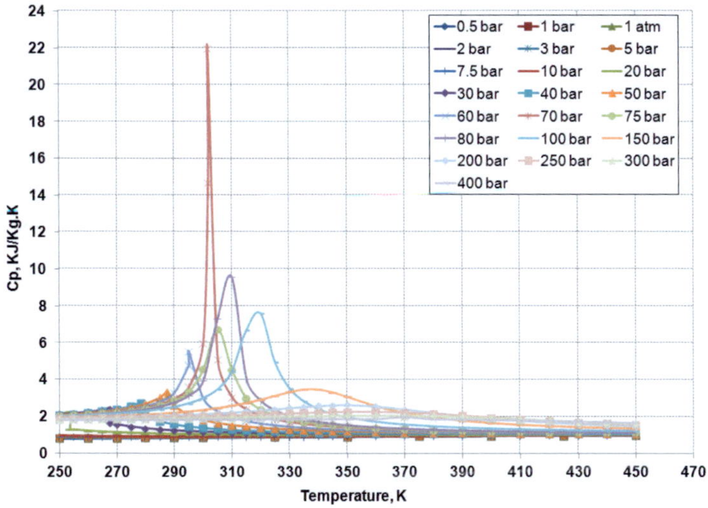

Figure B.4: Specific heat capacity at constant pressure vs. temperature in the thermodynamic properties' tables

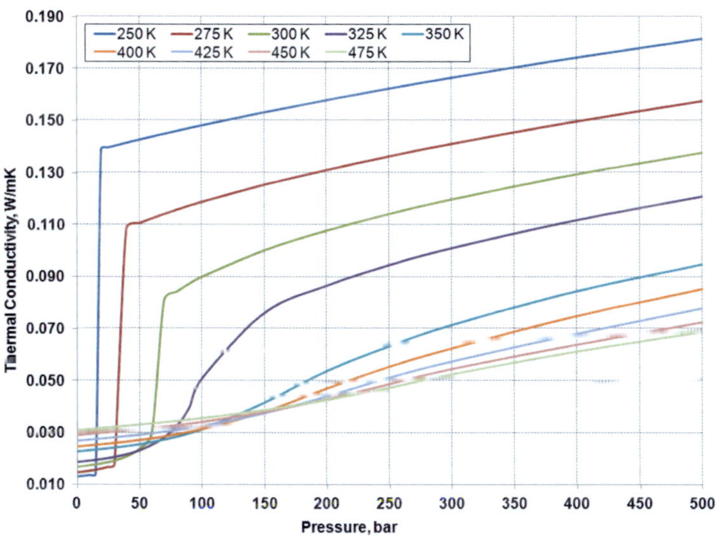

Figure B.5: Isotherms of thermal conductivity vs. pressure in the transport properties' tables

APPENDIX C - Simulation Results

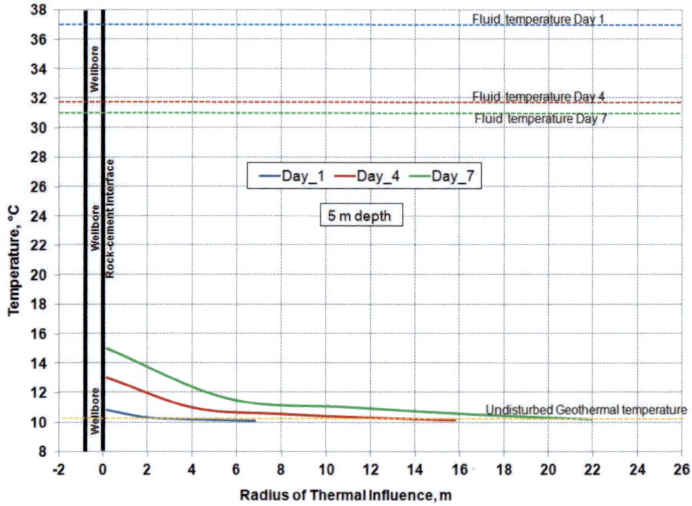

Figure C.1: *Prediction of temperature propagation in the radius of thermal influence in the top part of the Ketzin well*

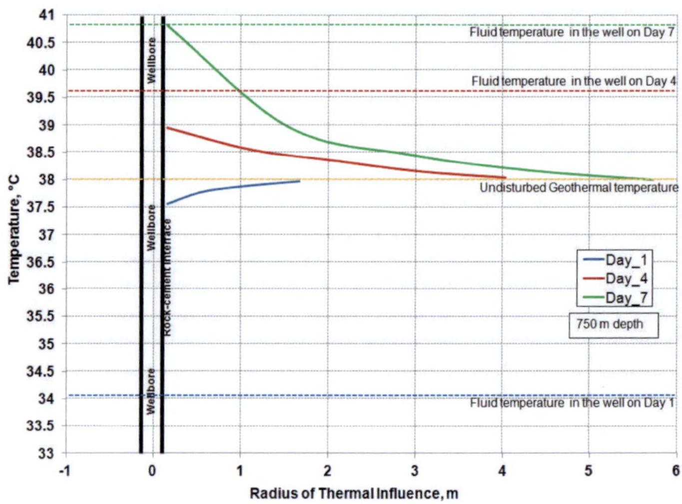

Figure C.2: *Prediction of temperature propagation in the radius of thermal influence in the bottom part of the Ketzin well*

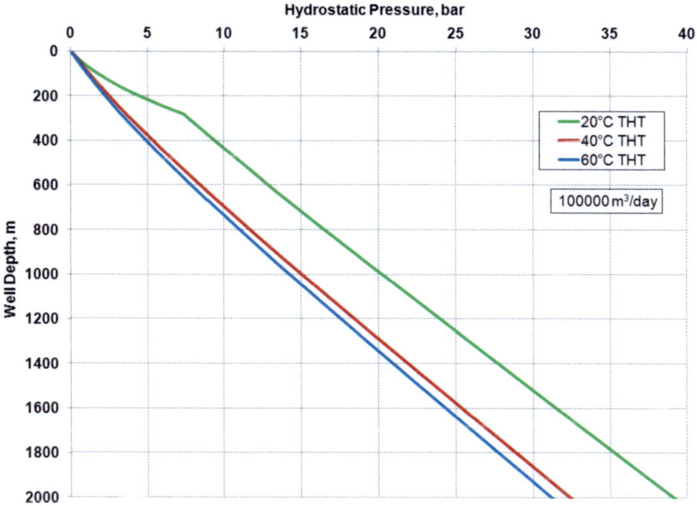

Figure C.3: *Sensitivity of hydrostatic pressure to wellhead temperature*

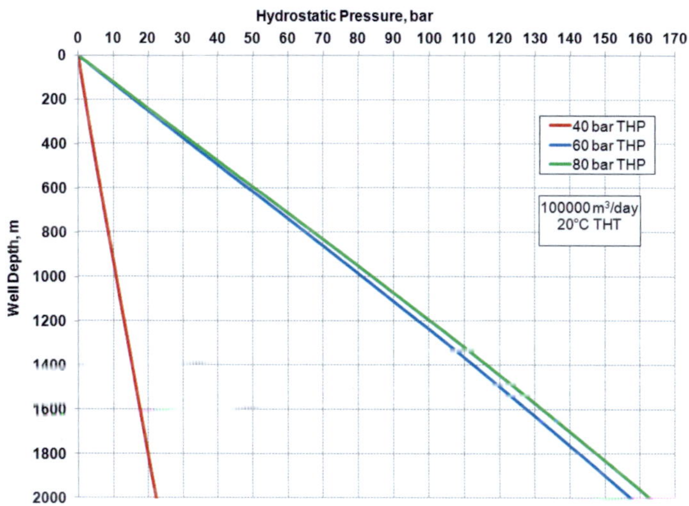

Figure C.4: *Sensitivity of hydrostatic pressure to wellhead pressure*

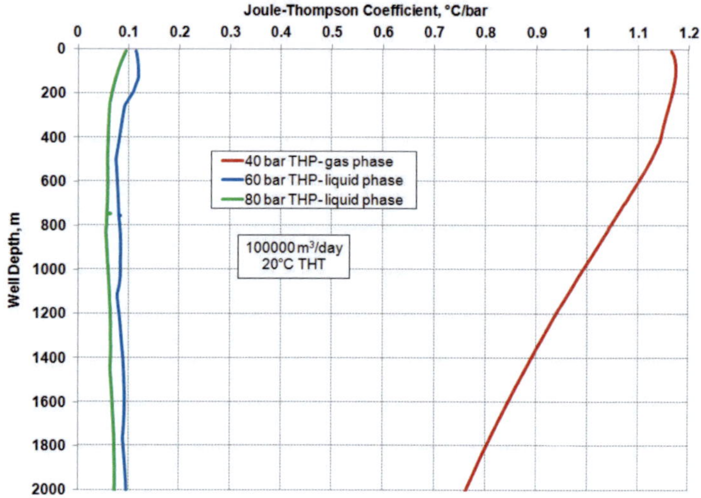

***Figure C.5**: Sensitivity of Joule-Thomson coefficients to wellhead pressure*

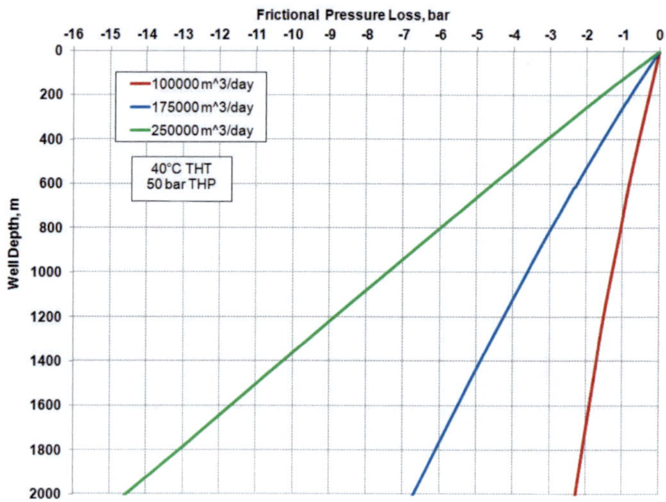

***Figure C.6**: Sensitivity of frictional pressure losses to flow rate*